HOLT, RINEHART AND WINSTON

NUMBER TOOLS

BRITANNICA Mathematics in Context

WORKBOOK

468 + 29

$468 \xrightarrow{+30} 498 \xrightarrow{-1} 497$

Sun City
Population: 625
Horses: 125
WANTED
REWARD

Moon City
Population: 120
Horses: 720

Miles	210	2100	700	100	25
Gallons of Gas	8.4	84	28	4	1

21,653 + 99 = 21,752

Britannica

Mathematics in Context is a comprehensive curriculum for the middle grades. It was developed in collaboration with the Wisconsin Center for Education Research, School of Education, University of Wisconsin–Madison, and the Freudenthal Institute at the University of Utrecht, The Netherlands, with the support of National Science Foundation Grant No. 9054928.

National Science Foundation

Opinions expressed are those of the authors
and not necessarily those of the Foundation

ISBN 0-03-072582-8

2 3 4 5 170 05 04 03

The *Mathematics in Context* Development Team

Mathematics in Context is a comprehensive curriculum for the middle grades. The National Science Foundation funded the National Center for Research in Mathematical Sciences Education at the University of Wisconsin–Madison to develop and field test the materials from 1991 through 1996. The Freudenthal Institute at the University of Utrecht in The Netherlands, as a subcontractor, collaborated with the University of Wisconsin–Madison on the development of the curriculum.

National Center for Research in Mathematical Sciences Education Staff

Thomas A. Romberg
Director

Joan Daniels Pedro
Assistant to the Director

Gail Burrill
Coordinator
Field Test Materials

Margaret Meyer
Coordinator
Pilot Test Materials

Mary Ann Fix
Editorial Coordinator

Sherian Foster
Editorial Coordinator

James A. Middleton
Pilot Test Coordinator

Project Staff

Jonathan Brendefur
Laura J. Brinker
James Browne
Jack Burrill
Rose Byrd
Peter Christiansen
Barbara Clarke
Doug Clarke
Beth R. Cole
Fae Dremock
Jasmina Milinkovic
Margaret A. Pligge
Mary C. Shafer
Julia A. Shew
Aaron N. Simon
Marvin Smith
Stephanie Z. Smith
Mary S. Spence

Freudenthal Institute Staff

Jan de Lange
Director

Els Feijs
Coordinator

Martin van Reeuwijk
Coordinator

Project Staff

Mieke Abels
Nina Boswinkel
Frans van Galen
Koeno Gravemeijer
Marja van den Heuvel-Panhuizen
Jan Auke de Jong
Vincent Jonker
Ronald Keijzer
Martin Kindt
Jansie Niehaus
Nanda Querelle
Anton Roodhardt
Leen Streefland
Adri Treffers
Monica Wijers
Astrid de Wild

Contents

BRITANNICA Mathematics in Context

Contents (continued)

BRITANNICA Mathematics in Context

Section A.
The Number Line

Name ______________________ Date ______________

Locating Numbers (Page 1 of 2)

1. Jason wants to place the ages of his family members in order.

Family Member	**Age** (years)
Jason	13
Mom	48
Dad	50
Sister Alicia	15
Aunt Paula	30
Grandfather	74
Great-grandmother	93

He uses the following number line, which extends from 0 to 100, because it includes all of the ages of his family members. Estimate where the family members' ages would be on the number line and write their names above the line.

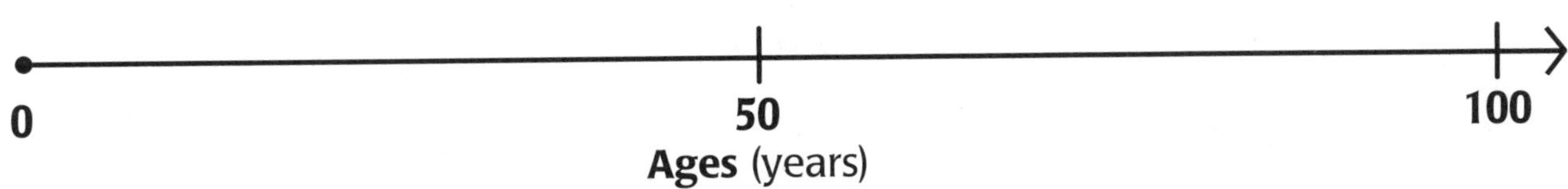

2. A group of students had a paper airplane contest. The Distance List below shows the number of centimeters each airplane flew. Locate the numbers for the flight distances on the number line. Reasonable estimates will do.

Distance List

Student	**Flight Distance** (cm)
Jim	244
Shantha	367
Lester	120
Giorgio	167
Bido	250
Alice	203
Arba	278
Martha	385

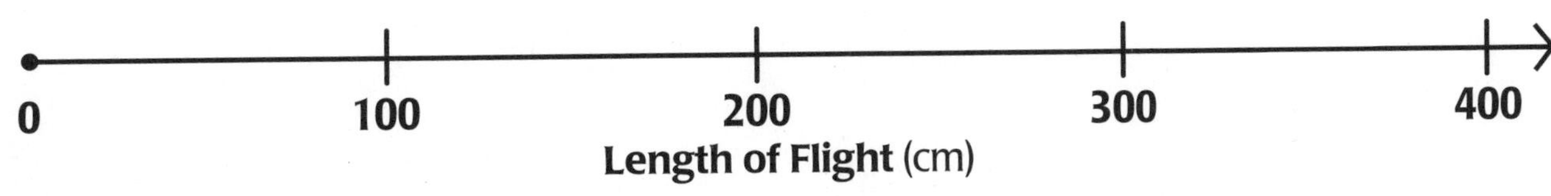

Name ______________________ Date ______________

Locating Numbers (Page 2 of 2)

The students then organized a second contest. The following graph shows the results of this contest.

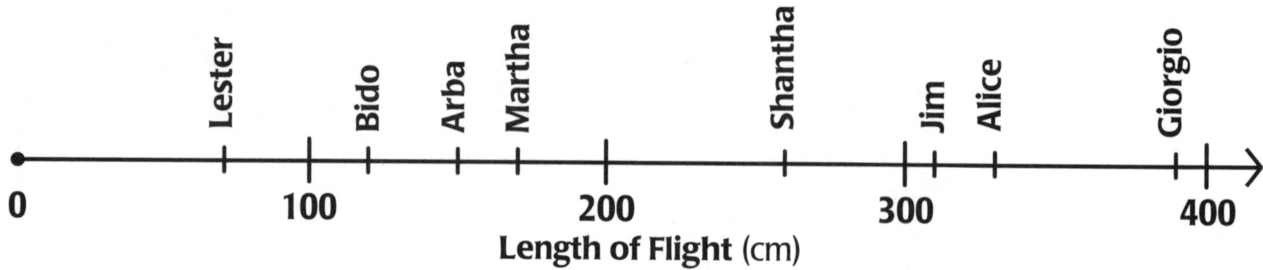

3. Complete the new Distance List by estimating each distance.

Distance List

Jim . . . . ________

Shantha . . . . ________

Lester . . . . ________

Giorgio . . . . ________

Bido . . . . ________

Alice . . . . ________

Arba . . . . ________

Martha . . . . ________

4. Compare your answers with those of a classmate. Are your answers exactly the same? Explain why or why not.

Name ______________________ Date ______________

The Jump, Jump Game (Page 1 of 3)

Object of the Game: To "jump" from one number to another on the number line in as few jumps as possible.

Number of Players: Two

To Play: Players should complete problems **1–10,** moving from one number to another in the fewest jumps possible.

To get to a number, players can make jumps of only three lengths: 1, 10, and 100. Players can jump forward or backward. Each player completes the problems separately.

After each round, compare the number of jumps each player took. The player with the fewest jumps should put an "x" in the box to the right of his or her number line.

If both players have the same number of jumps, then each player should put an "x" in the box.

An "x" is worth one point. The player with the most points after 10 rounds wins the game.

NOTE: Players can show their jumps on the number line by drawing curves of different lengths: a small curve for a jump length of 1, a medium curve for a jump length of 10, and a large curve for a jump length of 100.

Example: Go from 0 to 26 in the fewest jumps.

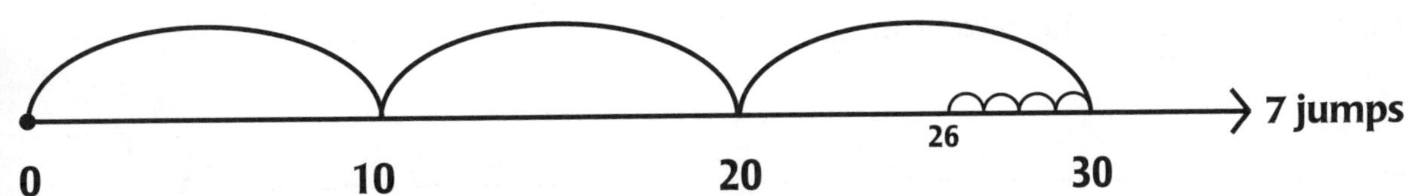

Name ______________________ Date ______________

Section A. The Number Line

The Jump, Jump Game (Page 2 of 3)

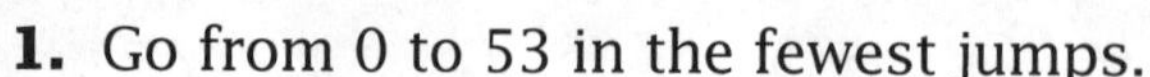

1. Go from 0 to 53 in the fewest jumps.

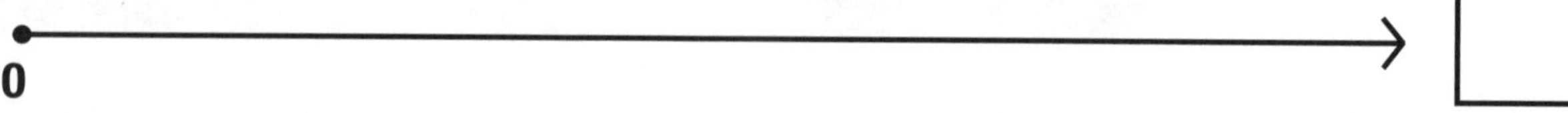

2. Go from 0 to 29 in the fewest jumps.

3. Go from 0 to 69 in the fewest jumps.

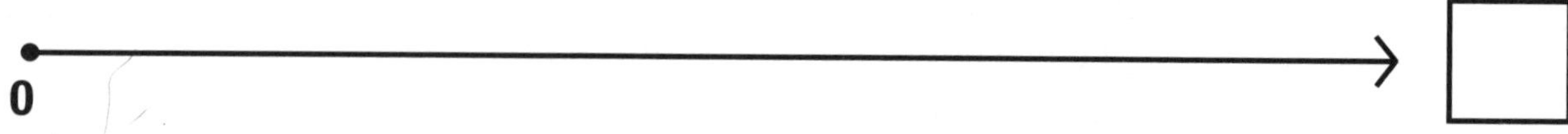

4. Go from 0 to 83 in the fewest jumps.

5. Go from 0 to 57 in the fewest jumps.

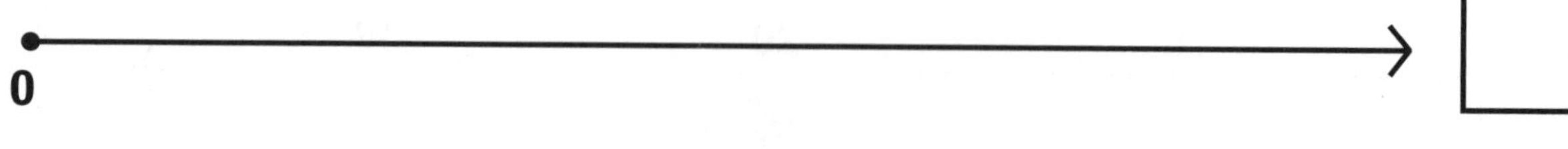

Name ______________________ Date ____________

The Jump, Jump Game (Page 3 of 3)

6. Go from 4 to 79 in the fewest jumps.

7. Go from 45 to 87 in the fewest jumps.

8. Go from 56 to 173 in the fewest jumps.

9. Go from 324 to 546 in the fewest jumps.

10. Go from 1492 to the current year in the fewest jumps.

Name ______________________ Date ______________

The Number Line as a Tool (Page 1 of 3)

Suppose you work at a booth selling beads. A customer wants to buy 29 beads of one kind and 67 beads of another. The two kinds of beads are the same price. How would you find the total number of beads? The number lines below show four strategies.

1. Study the number lines and explain the strategy shown on each. Then circle the letter of the strategy you would choose if you had to solve the problem *mentally*.

a.

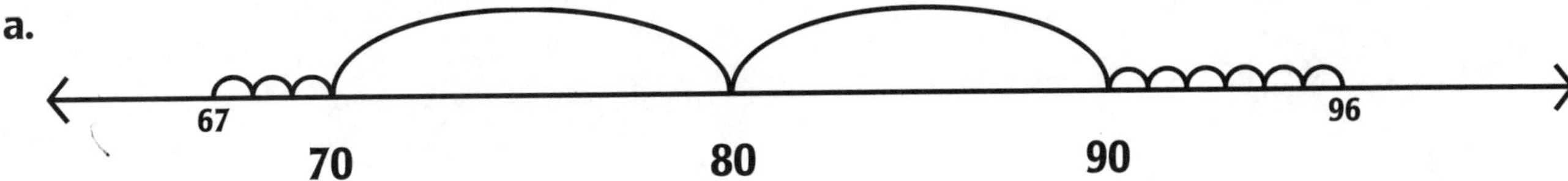

b.

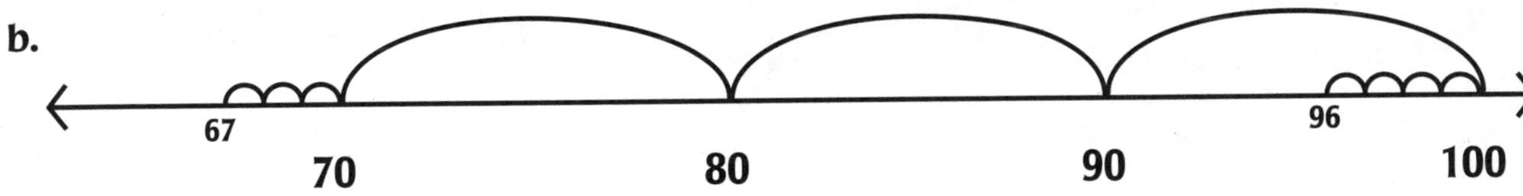

c.

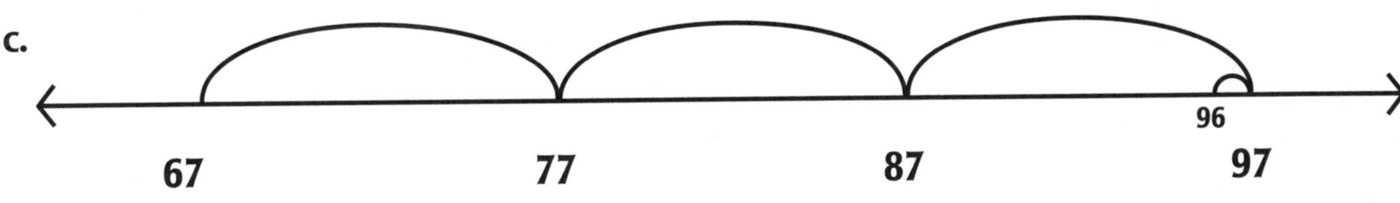

d.

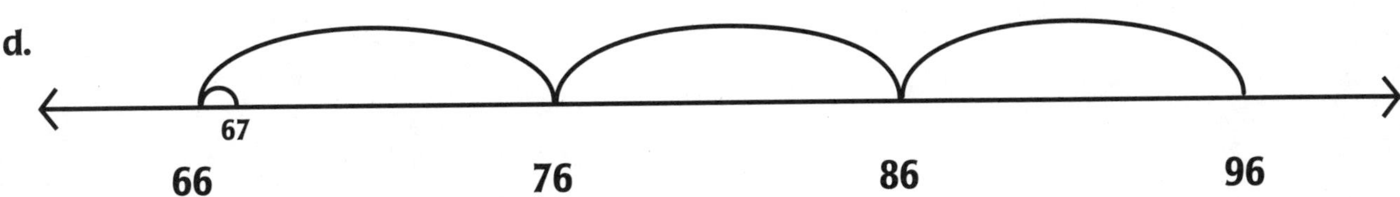

Name ______________________ Date ______________

The Number Line as a Tool (Page 2 of 3)

For problems 2–5, draw a number line that shows how you would solve each problem mentally. Explain your solution.

2. Mira's necklace has 18 beads. Nancy's necklace has 75 beads more than Mira's. How many beads does Nancy's necklace have?

3. The road from Red Bay to Buena Vista is 74 miles long.The road from Red Bay to College Park is 38 miles long. How much farther is it from Red Bay to Buena Vista than from Red Bay to College Park?

Name ______________________ Date ____________

The Number Line as a Tool (Page 3 of 3)

4. Ann's book has 364 pages. She has already read 158 pages. How many more pages does she have to read to finish the book?

5. Tim saved $376. Then he spent $99. How much money does he have left?

Name ______________________ Date ______________

Visualize the Number Line (Page 1 of 1)

Solve these addition problems *mentally.* It may be helpful to visualize the number line as you solve them.

1. José bought a pair of concert tickets for \$36. The ticket agency then sold him a third ticket for \$17. What was the total amount he spent on the tickets?

2. Minh needs to buy new wheels for her bicycle. The rear wheel costs \$56, and the front wheel costs \$38. How much will she have to pay for both wheels?

3. Marcel set a new school record by eating as many hot peppers as he could. The old school record was 46, and Marcel ate 19 more than that. How many hot peppers did he eat?

4. Ms. Bolivar shopped at two stores. At one store she bought \$35 worth of vegetables. At the other store, she bought \$66 worth of meat. What is the total amount she spent at both stores?

5. 69 + 38 =

6. 328 + 63 =

7. 956 + 98 =

Name ________________________ Date ____________

More Visualizing the Number Line (Page 1 of 1)

Try to solve these problems *mentally.* It may be helpful to visualize the number line as you solve them.

1. There are 53 fifth-grade students in one school. Eight of them go home with the flu. How many are still at school?

2. There were 84 cookies at a bake sale. Sixty-nine of them have been sold. How many are left?

3. Twenty-nine people drove their cars out of a school parking lot during the lunch hour. There had been 96 cars in the parking lot before lunch. How many cars remained in the lot after the 29 people left?

4. Vic's grandmother is 73 years old. She has lived in the same town for the past 65 years. How old was she when she moved there?

5. Qiana had some candy. She gave 27 pieces of candy to her friend Sharlonda, and now she has 17 pieces left. How many pieces of candy did she have to begin with?

6. 567 – 39 =

7. 972 – 53 =

8. 986 – 498 =

Name ________________________ Date ____________

Make Things Easy (Page 1 of 1)

Solve these problems *mentally.* It may be helpful to make a mental model of the number line.

1. Jonathan is having a party for a few of his closest friends. The bill comes to $249, and he wants to give the caterers a tip. He figures that $38 is a good tip. How much money does he give the caterers in total?

2. The local theater is showing a horror movie, and 674 people show up to watch it. If 25 people get scared and walk out, how many people will still be in the theater?

3. Malcolm X Jr. High School is holding a dance. If 441 girls and 423 boys show up, how many more girls than boys will there be?

4. The Yum Yum Cookie Factory adds 55 pounds of sugar to 777 pounds of cookie dough. How many pounds of cookie dough are there now?

5. At a track meet, Lisa jumped 652 centimeters in the long jump competition, and Shelice jumped 595 centimeters. How much farther did Lisa jump than Shelice?

6. The local auditorium has 1,002 seats. One night, 998 people show up for a concert. How many empty seats are there?

7. Create and solve your own problem using the numbers 508 and 470.

BRITANNICA

Mathematics in Context

Section B.
The Ratio Table

Summer Camp (Page 1 of 1)

Ms. Lampert organizes a summer camp for children. This year, 169 children have signed up for camp. Ms. Lampert needs to figure out how many tents are needed. Each tent is large enough for 12 children.

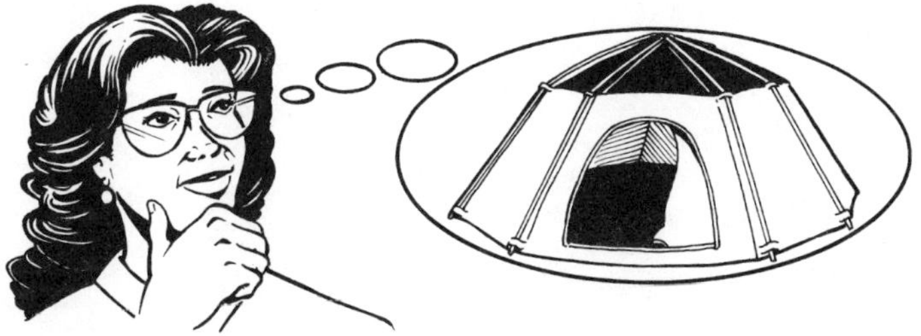

1. How many tents does Ms. Lampert need? Show at least two different ways that you can solve the problem *without using a calculator.*

Strategy 1

Strategy 2

This problem can be solved in many different ways. You might have written down all of the steps you took to find the answer. Using a ***ratio table*** is a good way to keep track of all of the steps.

With a ratio table, you start with a known ratio (in this case, 1 tent for 12 children) and use it to find other numbers with the same ratio (10 tents for 120 children, for instance). You can keep using the numbers you find to discover more numbers, until you solve the problem. For instance, look at the process used in the table below.

Tents	1	10	5	15	14
Children	12	120	60	180	168

The second column of numbers is the first column times 10. The third column is half of the second column. The fourth column is the total of the second and third columns, and the fifth column is the fourth column minus the first column. Then you can figure that if 14 tents will hold 168 children, 15 tents will be needed to hold 169.

2. Compare your original two strategies with the strategy of using a ratio table. Which strategy do you like the best? Why?

Name ______________________ Date ______________

Bottles (Page 1 of 2)

1. Camp's Mineral Water is shipped in cases. Trudy Camp, owner of the company, uses the table below to keep track of the number of cases she has to ship each time a customer orders a certain number of bottles. Each case holds 15 bottles. Find out how many bottles there are in two, three, four, or more cases. Write your answers in the table below.

Cases	1	2	3	4	5	6	7	8
Bottles	15							

2. Jake's convenience store orders 90 bottles of mineral water. How many cases should be shipped?

3. Jake orders 65 bottles of mineral water for a work party. How many cases is this?

4. Camp's Seltzer comes in a larger case that holds more than 15 bottles. Trudy Camp's table for seltzer was ruined when she spilled water on it. The table below shows the few numbers that Trudy can still read. Can you tell from the table how many bottles there are in each case? Explain your answer.

New Cases					5	6						12
Bottles					125	150						300

5. How many bottles would fit in 18 of the larger seltzer cases?

Name ____________________ Date ____________

Bottles (Page 2 of 2)

6. Camp's juice cases hold only 14 bottles each. Use the following ratio tables to find out how many bottles there are in different numbers of cases:

a. 8 cases

Cases	1	2	4	8
Bottles	14			

b. 6 cases

Cases	1	2	3	6
Bottles	14			

c. 5 cases

Cases	1	10	5
Bottles	14		

d. 9 cases

Cases	1	10	9
Bottles	14		

7. Jake orders 98 bottles of juice from Camp's. Use the following ratio table to determine how many cases will be shipped. Add more columns if necessary.

Cases	1				
Bottles	14				

Name ____________________ Date ____________

Plants (Page 1 of 2)

Mr. Martin's biology class is starting a school garden. They will be ordering plants by the box from a nursery. Mr. Martin asked the class to figure out how many tomato plants are in 16 boxes if one box contains 35 plants.

Three students—Darrell, Tasha, and Carla—solved the problem using ratio tables, but each student used a different table.

1. Darrell solved the problem as shown below:

Boxes	1	2	3	4	5	6	7	8	16
Plants	35	70	105	140	175	210	245	280	560

Explain Darrell's solution.

2. Tasha solved the problem as shown below:

Boxes	1	2	4	8	16
Plants	35	70	140	280	560

Explain Tasha's solution.

3. Carla solved the problem as shown below:

Boxes	1	10	2	6	16
Plants	35	350	70	210	560

Explain Carla's solution.

Name ______________________ Date ______________

Plants (Page 2 of 2)

4. Think of another way to use a ratio table to figure out how many tomato plants are in 16 boxes. Show your solution in the table below. You may add as many columns as you need to the table.

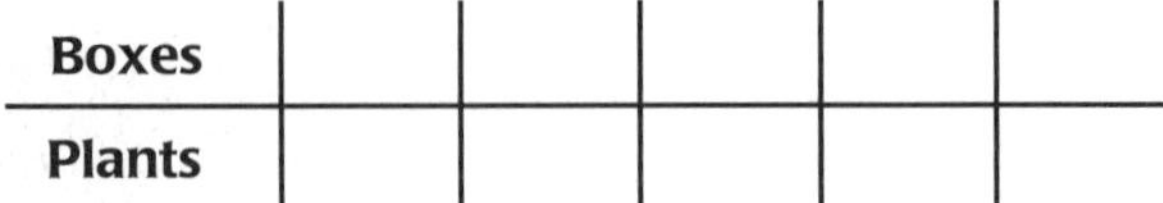

Boxes					
Plants					

5. Cactuses are shipped 45 pots to a box. If Mr. Martin's class orders 360 cactuses, how many boxes will they get? Show your strategy using the following ratio table.

Boxes					
Cactuses					

6. Mr. Martin's students decide that they need 675 cactuses. How many boxes will arrive?

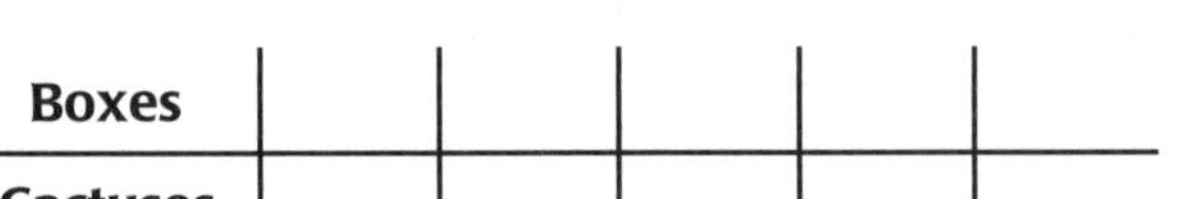

Boxes					
Cactuses					

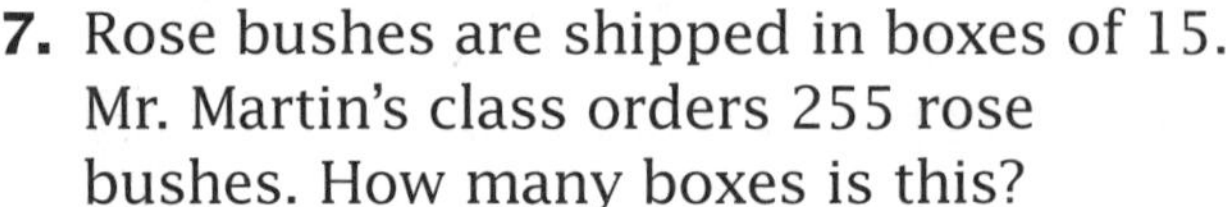

7. Rose bushes are shipped in boxes of 15. Mr. Martin's class orders 255 rose bushes. How many boxes is this?

Boxes					
Roses					

8. Strawberry plants are shipped in boxes of 70. If Mr. Martin's class orders 980 strawberry plants, how many boxes will arrive?

Boxes					
Strawberry Plants					

9. a. Are problems **6–8** multiplication or division problems? Explain your answer.

b. Does your answer to part **a,** above, make any difference in the way you would solve problems **6–8?**

Name ______________________ Date ______________

Going to the Movies (Page 1 of 2)

1. Mary Lou has a new job, selling tickets at a movie theater. A ticket costs $5.50, but many people order two or more tickets. Mary Lou decides to make a table to help her determine the total price more quickly. Fill in the prices for the different numbers of tickets below.

Tickets	1	2	3	4	5	6	7	8	9	10
Dollars	5.50									

2. Sing wanted to go to the movies with five friends. Since she lives near the theater, Sing offered to buy the tickets for everyone in advance. Sing was not sure if she had enough money to buy all six tickets. The table below shows the method she used to calculate the amount of money she needed. Explain Sing's way of calculating the answer.

Tickets	1	2	4	6
Dollars	5.50	11.00	22.00	33.00

3. Otacilio has to buy 12 tickets for a play. One ticket costs $14. Calculate the price for 12 tickets using the following ratio table.

Tickets	1				
Dollars	14.00				

4. Kaleen wants to take eight children from her neighborhood to a theater performance; the tickets cost $12 each. How much must Kaleen pay for nine tickets? Use the following ratio table to find your solution.

Tickets	1			
Dollars	12.00			

Name ______________________________ Date ______________

Going to the Movies (Page 2 of 2)

5. One second-grade class wants to go to the theater. If tickets are priced at $16 each, what do 25 tickets cost? Use the following ratio table to find your solution.

Tickets	1			
Dollars	16.00			

6. Mary Lou needs to sell 16 tickets to a large group, but her chart does not go up to 16. Knowing that the cost of one ticket is $5.50, Mary Lou calculates the cost of 16 tickets like this:

Tickets	1	10	2	4	6	16
Dollars	5.50	55.00	11.00	22.00	33.00	88.00

Explain how Mary Lou solved the problem.

7. Leroy, who works with Mary Lou, solved the same problem using this method:

Tickets	1	2	4	8	16
Dollars	5.50	11.00	22.00	44.00	88.00

Explain Leroy's solution.

8. Leah was asked to calculate the total price of 15 tickets for a special documentary film. Each ticket cost $6. This is the way she did it:

Tickets	1	10	5	15
Dollars	6.00	60.00	30.00	90.00

Explain Leah's method.

Name ______________________ Date ______________

Fruits (Page 1 of 1)

In order to solve the problems on this page, use the ratio table given with each problem. If necessary, add extra columns.

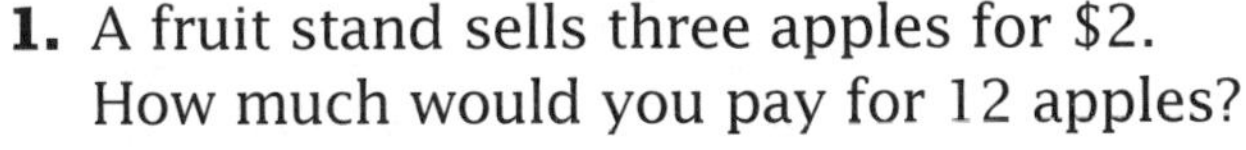

1. A fruit stand sells three apples for $2. How much would you pay for 12 apples?

Apples	3			
Dollars	2.00			

2. What would you pay for 33 apples? Choose your own method to calculate the answer.

Apples	3			
Dollars	2.00			

3. The fruit stand sells five big, juicy oranges for $3.75. If you needed 35 oranges, how much would you have to pay?

Oranges	5			
Dollars	3.75			

4. Another type of orange is sold four for $2.50. You have $10 to spend. How many oranges can you buy?

Oranges	4			
Dollars	2.50			

5. Cantaloupes are sold two for $3. How many cantaloupes can you buy for $7.50?

Cantaloupes	2			
Dollars	3.00			

Name ______________________ Date ______________

Section B. The Ratio Table

How Do You Do It? (Page 1 of 2)

Using ratio tables is a convenient way to solve some problems. For instance, you can use ratio tables to find:

- how many cases you need for different numbers of bottles,
- how much different numbers of tickets cost,
- the amount of fruit you can buy with a given amount of money.

You may have discovered that the ratio table offers a handy way to write down intermediate steps you take to solve a problem.

There are several ways to use existing numbers to find new numbers:

adding

1	2	3
35	70	105

times 10

1	10
35	350

doubling

1	2	4	8
35	70	140	280

halving

1	10	5
28	280	140

subtracting

1	10	9
15	150	135

multiplying

2	6
70	210

Find the missing number in each of the following problems. Then think about how you did it. Circle the numbers you used to find the answer and the word that best describes your approach.

1. In Tanisha's shop, 1 meter of yellow curtain fabric costs $12. How much would 9 meters of the same fabric cost?

Meters	1	10	9
Dollars	12	120	

adding • times 10 • doubling
halving • subtracting
multiplying

2. It costs $28 per day to rent a sailboat at the Harbor Club. How much would it cost to rent a sailboat for five days?

Days	1	10	5
Dollars	28	280	

adding • times 10 • doubling
halving • subtracting
multiplying

Name ______________________ Date ______________

How Do You Do It? (Page 2 of 2)

3. Anton earns $13 a day at his summer job, baby-sitting his cousins. How much will he earn in 20 days?

Days	1	10	20
Dollars	13	130	

adding • times 10 • doubling
halving • subtracting
multiplying

4. One bottle of apple juice costs $2.75. How much do four bottles of apple juice cost?

Bottles	1	2	4
Dollars	2.75	5.50	

adding • times 10 • doubling
halving • subtracting
multiplying

5. Mr. Pink wants to buy some stamps for his office. How much must he pay for 40 stamps if each stamp is worth 32¢?

Stamps	1	2	4	40
Dollars	0.32	0.64	1.28	

adding • times 10 • doubling
halving • subtracting
multiplying

6. Altagracia can fill six glasses with one bottle of apple juice. How many glasses can she fill with four bottles?

Bottles	1	4
Glasses	6	

adding • times 10 • doubling
halving • subtracting
multiplying

7. There are 24 bottles in a case. How many bottles are in nine cases?

Cases	1	2	4	10	5	9
Bottles	24	48	96	240	120	

adding • times 10 • doubling
halving • subtracting
multiplying

Name ______________________ Date ______________

Egyptian Arithmetic (Page 1 of 1)

The Egyptians drew hieroglyphs—pictures or symbols that have meanings that are understood by a group of people—on the walls of temples and tombs. They also used symbols to write numbers. Here are a few of their number symbols:

ONE

TEN

HUNDRED

THOUSAND

10 THOUSAND

100 THOUSAND

1 MILLION

10 MILLION

The Egyptian transcription of 1,324 would be:

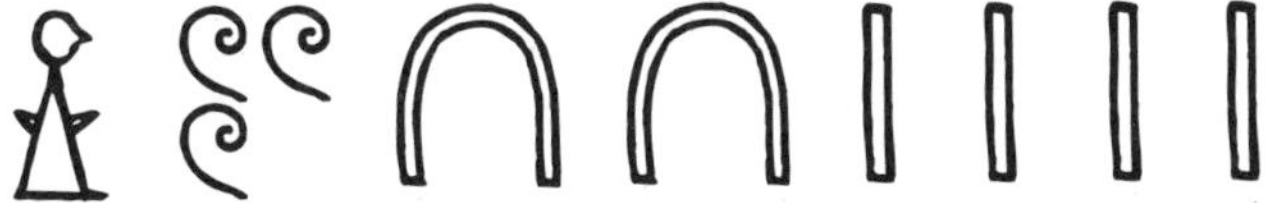

1. What number is being represented below?

2. How is the Egyptian method similar to our modern method of writing numbers? How are the two methods different?

3. What is the outcome—in Egyptian script—if you add the two numbers below? Write your answer using the fewest symbols possible.

4. What number represents the answer to problem **3?**

Name ______________________ Date ______________

Egyptian Multiplication and Division (Page 1 of 2)

The Egyptian way to solve multiplication and division problems resembles the way you have been solving multiplication and division problems in this unit, by using a ratio table. When solving a problem, the Egyptians wrote down intermediate steps just as you did. In their arithmetic, doubling was very important.

1. For example, this is how the Egyptians might have calculated 23 × 49 (except for the fact that we are using numbers). The bold numbers are used to find the outcome.

1	**49**
2	**98**
4	**196**
8	392
16	**784**
23	1,127

   a. What is the outcome of 1 + 2 + 4 + 16?

   b. What is the outcome of 49 + 98 + 196 + 784?

   c. Calculate 23 × 49 with your calculator.

2. Use the Egyptian method of repeated doubling to calculate 22 × 35.

Division problems can be solved in a similar way. If you want to know how many times six goes into 78, you can answer the multiplication question: Six times what equals 78?

1	**6**
2	12
4	**24**
8	**48**
13	78

6 + 24 + 48 = 78

1 + 4 + 8 = 13

6 × 13 = 78 so 78 ÷ 6 = 13

3. Most years contain 365 days. Calculate, by repeated doubling, approximately how many weeks there are in a year. Does it come out evenly? Explain.

Name ______________________ Date ______________

Egyptian Multiplication and Division (Page 2 of 2)

4. The Egyptians discovered that the outcome of "something times 10" is easy to find. Write down, in Egyptian symbols, the outcome of 10 times the number shown below. Explain how you got your answer.

5. Because our number system is based on 10, the outcome of "10 times something" is easy to find. The Egyptian number system was also based on 10. When you compare the way they had to find the outcome of 10 × 35 with the way we do it, which system do you prefer? Why?

6. **a.** Solve 78 ÷ 6 using a ratio table. Try to use a method other than doubling.

   b. Do you prefer the Egyptian method of doubling to find 78 ÷ 6, as was done on the previous page, or the method you just used for problem **6a?** Why?

BRITANNICA Mathematics in Context

Section C. Number Sense

Name ______________________ Date ______________

What's in a Number? (Page 1 of 1)

How could you describe the number 52? You might say:

52 is 2×26;
52 is 4×13;
52 is two quarters and two pennies; or
52 is the age of Anton's father.

1. Write down several other ways to describe the number 52.

2. Write down several ways to describe the number 64.

Name ______________________ Date ______________

Door Prizes (Page 1 of 1)

Keesha and Anthony are arranging door prizes for a school party. They decide to make packages of items whose total value is $1. The items Keesha and Anthony can use are shown below. Two of the packages have already been put together. Find at least 10 more package groupings for Keesha and Anthony.

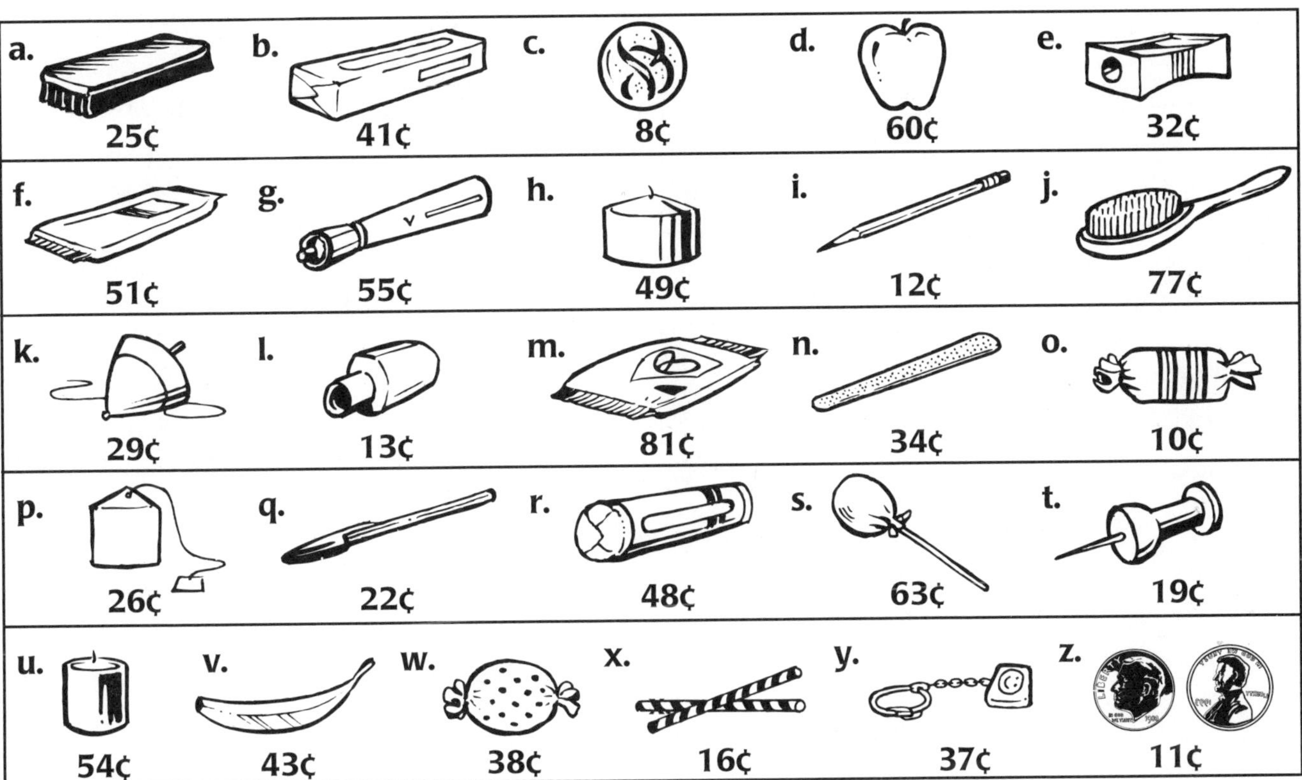

Door Prize Groupings

1. t and m (19¢ and 81¢)

2. 4 i's and 4 l's
(12¢ × 4 and 13¢ × 4, or 48¢ and 52¢)

3.

4.

5.

6.

7.

8.

9.

10.

11.

12.

Name ______________________ Date ______________

Ticket Totals (Page 1 of 1)

Mr. Tchang's class is participating in a walkathon for charity. Each student found sponsors who will donate money according to how far the student walks. Mr. Tchang asked the students each to write down how many sponsors he or she has. Below are the pieces of paper the students gave Mr. Tchang.

Figure out how many total sponsors Mr. Tchang's class has. Because there are so many numbers, be sure to use an adding strategy to find the sum. Explain your strategy.

Name ______________________ Date ______________

Unknown Total (Page 1 of 1)

Don has the flu and cannot leave his house. He asked his friend Ebony to buy him some items at the local grocery store. Ebony bought the groceries on his list and saved the receipt, but the total was torn off when the cashier took the receipt from the cash register.

CASH RECEIPT
STORE # 2204

.99
.78
.74
.78
.24
.15
.34
.67
.63
.25
.54
.90
.35
.82

TOTAL $

1. How can the numbers be combined to help Don and Ebony figure out the total?

2. Find the total amount without using your calculator.

Name ____________________ Date ____________

Small Change (Page 1 of 1)

Carmen wanted to buy some gifts for a holiday party. She shopped in different stores to find certain items.

Carmen needed a quarter for the parking meter, so at the first store, she paid $1.10 for an 85¢ item.

Price of Item Bought	Amount Given to Cashier	Change
$ 0.85		

1. Below are four more transactions. The total prices and the change given back to Carmen are shown. Find the amount Carmen gave the cashier for each transaction.

	Price of Item Bought	Amount Given to Cashier	Change
a.	$ 0.96		
b.	$ 0.67		
c.	$ 0.55		
d.	$ 0.89		

2. Make up your own prices, amounts given to the cashier, and change received.

	Price of Item Bought	Amount Given to Cashier	Change
a.	$		
b.	$		
c.	$		
d.	$		

Name ____________________ Date ____________

The Vending Machine (Page 1 of 1)

How many quarters, dimes, and/or nickels do you have to put in the vending machine to get the following items? List all possible combinations. Three possibilities are listed for **item a.**

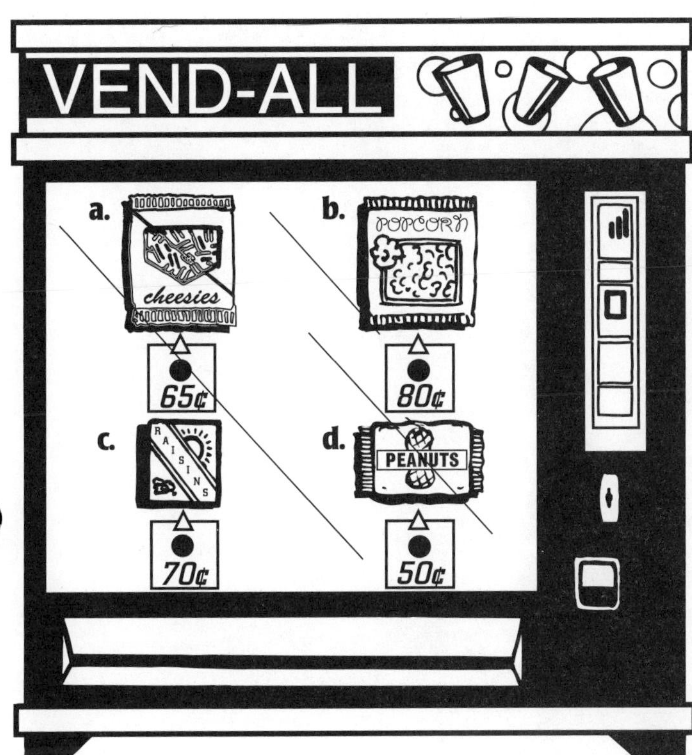

	Quarters	Dimes	Nickels
a.	2 1 1	1 4 3	1 0 2
b.			
c.			
d.			

The Right Price Game (Page 1 of 1)

Dorian watches his favorite game show, *The Right Price*, every day after school. Sometimes the contestants are shown three items. The host of the show tells the total price for the three items and the individual price of one of the items. In 15 seconds, the contestants must give as many possible prices for the other two items as they can. The more prices the contestants can give, the more prizes they win. But the three prices must always add up to the given total.

Now it is your turn. Play the following four games. Find as many prices as you can that will work for each game. But, just like the contestants on *The Right Price*, you may not use a calculator.

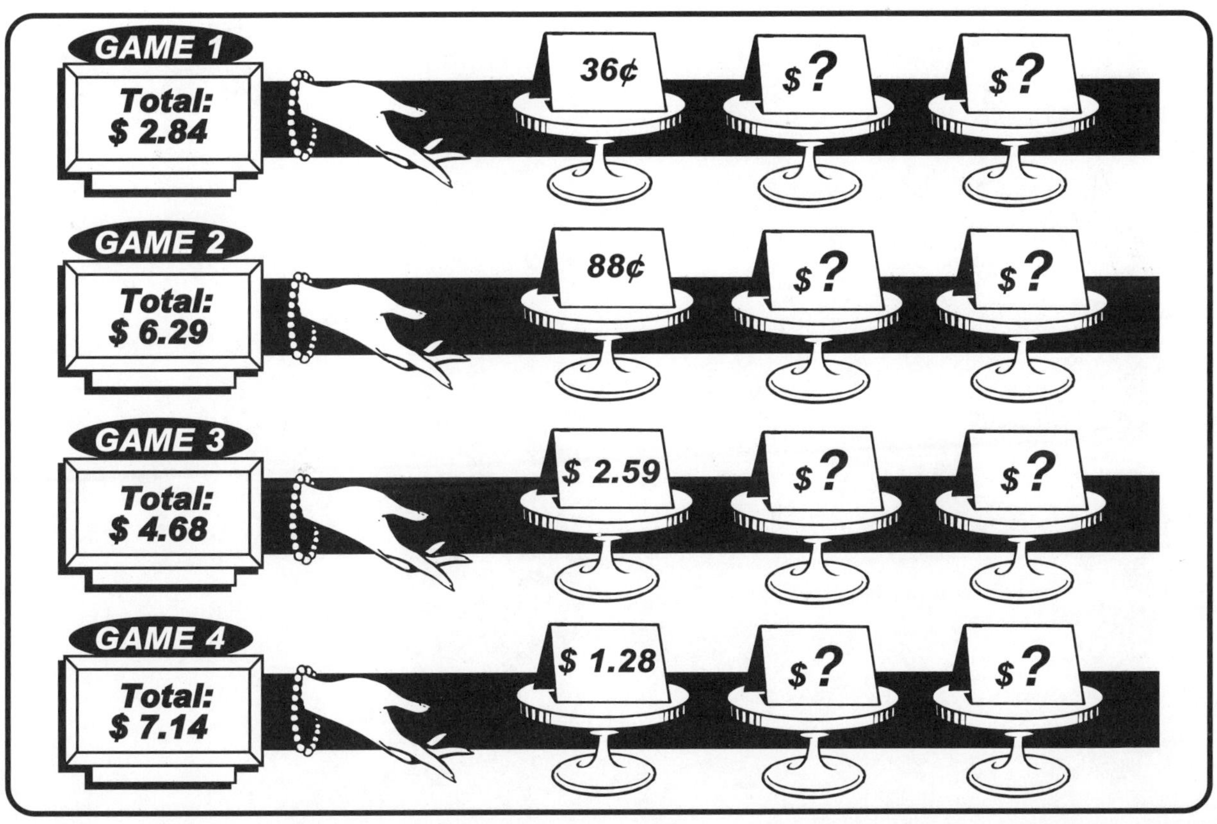

Name ______________________ Date ____________

The Plus and Minus Game (Page 1 of 1)

Object of the game: To reach a given number by adding and/or subtracting certain numbers in the fewest steps possible.

Number of players: 2 to 30

To play:

- Each player must add and/or subtract the numbers in the boxes on the left to reach the number in the box on the right. Players must use all of the numbers at least once, but may use some numbers more than once.
- The winner is the player who reaches the number in the fewest steps.

Example:

3	5	9	10	25

Judy used 3 + 5 + 9 + 10 + 3 − 5 = 25.

Meriza figured 3 + 5 + 9 + 10 + 10 − 9 − 3 = 25.

Since Judy used only five steps and Meriza used six, Judy won.

Play the game with the following numbers:

1. | 3 | 6 | 7 | 8 | 39 |

2. | 3 | 5 | 4 | 9 | 32 |

3. | 17 | 20 | 8 | 10 | 74 |

4. | 5 | 11 | 16 | 4 | 53 |

5. | 15 | 8 | 14 | 20 | 91 |

6. | 13 | 9 | 15 | 12 | 68 |

Name ______________________________ Date ______________

Working at the Post Office (Page 1 of 1)

A postal worker tears off the stamps she sells from large sheets. Below is part of a sheet of 6-cent stamps.

6¢	6¢	6¢	6¢	6¢	6¢	6¢	6¢	6¢	6¢
6¢	6¢	6¢	6¢	6¢	6¢	6¢	6¢	6¢	6¢
6¢	6¢	6¢	6¢	6¢	6¢	6¢	6¢	6¢	6¢
6¢	6¢	6¢	6¢	6¢	6¢	6¢	6¢	6¢	6¢
6¢	6¢	6¢							

1. a. How much is this part of the sheet worth? Explain how you found your answer.

b. One postal worker multiplied two numbers to find the total value of the stamps in the above sheet. What numbers did she multiply?

2. Imagine you are working at the post office. People ask for a number of stamps and you have to calculate how much they owe. What is the cost of each of the following sets of stamps? Explain how you calculated your answers.

a.

8¢	8¢	8¢	8¢	8¢
8¢	8¢	8¢	8¢	8¢
8¢	8¢			

b.

16¢	16¢	16¢	16¢	16¢
16¢	16¢	16¢	16¢	16¢
16¢	16¢	16¢	16¢	16¢

c.

25¢	25¢	25¢	25¢	25¢
25¢				

d.

50¢	50¢	50¢	50¢	50¢
50¢	50¢			

e.

40¢	40¢	40¢

f.

80¢	80¢	80¢	80¢	80¢
80¢	80¢	80¢		

g.

$1.20	$1.20	$1.20

Name ______________________ Date ______________

Stamp Costs (Page 1 of 1)

1. Below are sets of stamps that people bought at the post office. How much did each person have to pay?

a.

4¢	4¢	4¢	4¢	4¢
4¢	4¢	4¢	4¢	4¢
4¢	4¢	4¢	4¢	4¢

b.

9¢	9¢	9¢	9¢	9¢
9¢	9¢	9¢	9¢	9¢
9¢				

c.

$2.00	$2.00	$2.00	$2.00	$2.00
$2.00	$2.00	$2.00		

d.

25¢	25¢	25¢	25¢
25¢	25¢	25¢	25¢

e.

50¢	50¢	50¢	50¢
50¢	50¢	50¢	50¢

f.

75¢	75¢	75¢	75¢
75¢	75¢	75¢	75¢

2. For sheets of stamps that contain 10 rows of 10 stamps each, compute the costs using the following stamps.

   a. How much would a sheet of 32–cent stamps cost?

   b. How much would a sheet of 25–cent stamps cost?

   c. How much would a sheet of 48–cent stamps cost?

Name ______________________ Date ______________

Estimating Stamp Costs (Page 1 of 1)

When Mr. Lee buys stamps at the post office, he always pays with dollar bills because he never carries change. Estimate how many dollars Mr. Lee would need for each of the following sets of stamps.

1. 29¢ 29¢ 29¢ 29¢ 29¢ 29¢

2. 39¢ 39¢ 39¢ 39¢ 39¢ 39¢ 39¢ 39¢ 39¢ 39¢

3. 48¢ 48¢ 48¢ 48¢ 48¢ 48¢ 48¢ 48¢

4. 31¢ 31¢ 31¢ 31¢ 31¢ 31¢ 31¢

5. 52¢ 52¢ 52¢ 52¢ 52¢ 52¢ 52¢ 52¢

6. 58¢ 58¢ 58¢ 58¢

7. 97¢ 97¢ 97¢ 97¢ 97¢ 97¢ 97¢ 97¢ 97¢

8. 71¢ 71¢ 71¢ 71¢ 71¢ 71¢ 71¢

Name ____________________ Date ____________

Postage (Page 1 of 1)

Salina is going to mail some packages that each require $3.00 worth of postage. She checks her supply of stamps and discovers that she has several stamps worth 10 cents, 20 cents, and 30 cents.

1. a. Write down several different ways in which she could combine the stamps to equal the $3.00 postage.

b. How could she combine the stamps to equal $3.00 using the fewest stamps?

2. Salina also has to mail some packages that require $5.00 worth of postage.

a. Write down several different ways in which she could combine the 10-cent, 20-cent, and 30-cent stamps to equal the $5.00 postage.

b. How could she combine the stamps to equal $5.00 using the fewest stamps?

3. What is the total value of each of the following groups of stamps?

a.

14¢	14¢	14¢	14¢	14¢	14¢

b.

10¢	10¢	10¢	10¢	10¢	10¢
4¢	4¢	4¢	4¢	4¢	4¢

c.

20¢	20¢	20¢	20¢	20¢	20¢

d.

19¢	19¢	19¢	19¢	19¢	19¢

e.

10¢	10¢	10¢	10¢	10¢	10¢
9¢	9¢	9¢	9¢	9¢	9¢

f.

29¢	29¢	29¢	29¢	29¢	29¢

Name ______________________ Date ______________

Letter Postage (Page 1 of 2)

1. What is the total postage on four letters if each letter has 46 cents worth of stamps?

2. Chaska has to mail six letters, each requiring 34 cents in postage. How much will it cost him to mail these letters?

3. Jim has eight letters to mail, and each requires 66¢ worth of stamps. How much will it cost him to mail these letters?

4. How much will it cost to mail seven letters that each require 23 cents in postage?

5. How much will it cost to mail eight letters that each require 60¢ in postage?

6. What would it cost to mail nine letters if each had a postage requirement of 52¢?

7. Twelve letters have a postage rate of 21 cents each. How much will it cost to mail all 12 letters?

Name ______________________________ Date ______________

Letter Postage (Page 2 of 2)

8. This block of identical stamps costs $8. The value of one stamp is __________.

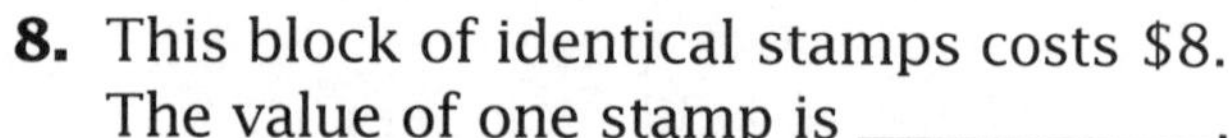

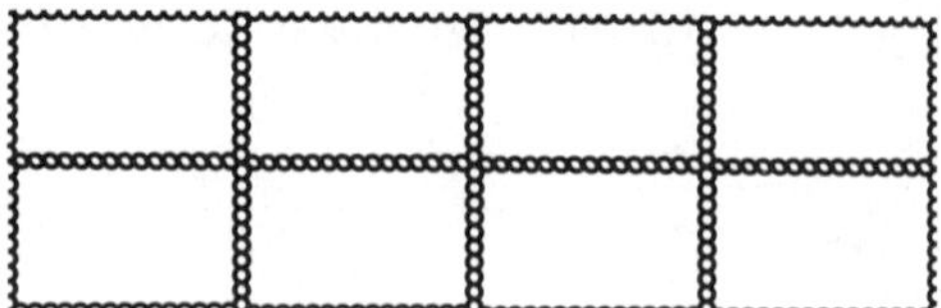

9. This block of identical stamps costs $2.40. The value of one stamp is ______.

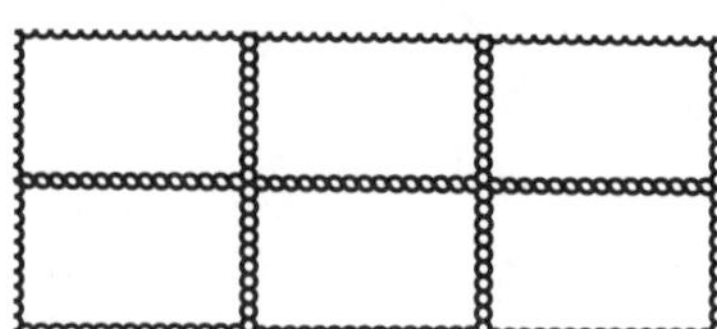

10. This block of identical stamps costs $2.70. The value of one stamp is ______.

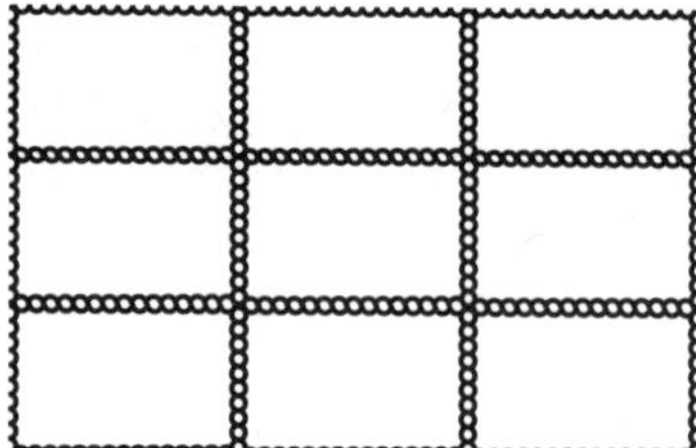

11. This block of identical stamps costs $4.50. The value of one stamp is ______.

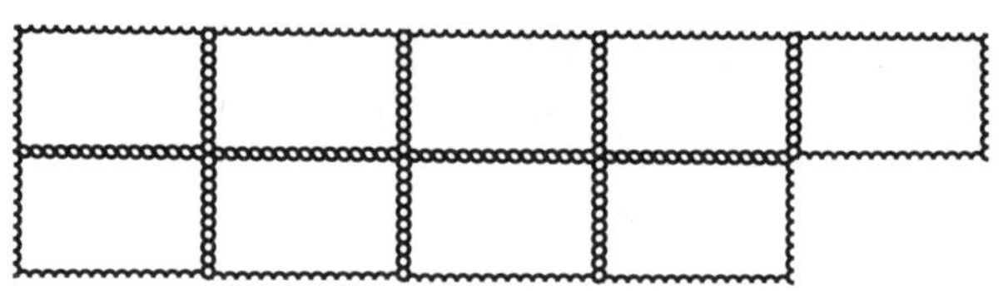

12. This block of identical stamps costs $3. The value of one stamp is __________.

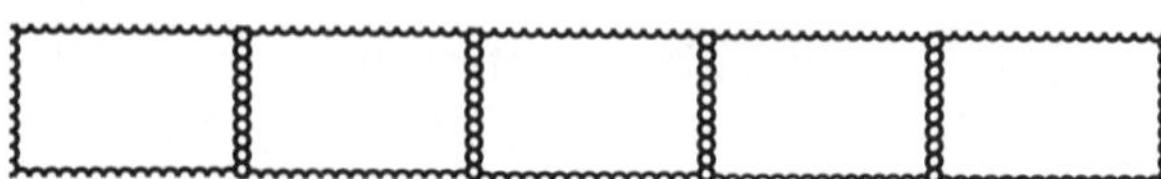

13. This block of identical stamps costs $24. The value of one stamp is __________.

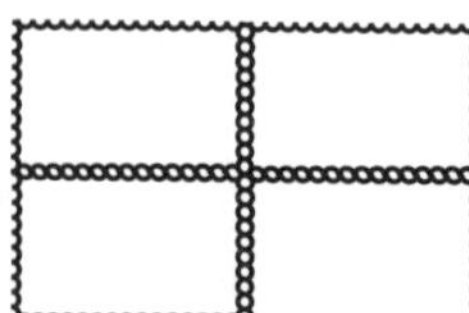

Name ______________________ Date ____________

Wrapping Paper (Page 1 of 2)

Liz and her sister Claire are going to their grandmother's birthday party. They have been riding in the car for six hours already. Their mother wants to give them something to do, so she asks who can count the figures on the wrapping paper on one side of each present the fastest.

1. a. How many birds are there on this section of wrapping paper?

b. What multiplication is appropriate? _____ × _____ = _____

2. a. How many palm trees are there on this section of wrapping paper?

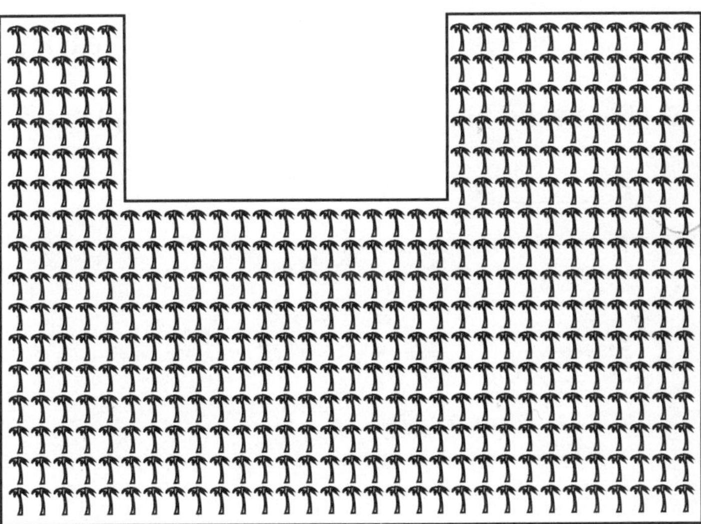

b. Explain the strategy you used for counting the palm trees.

Name ____________________ Date ____________

Wrapping Paper (Page 2 of 2)

On this small section of wrapping paper, there are three rows, each row containing eight roses, which gives a total of 24 roses.

3. How could you cut a piece of this type of wrapping paper so that it would contain exactly 120 roses? Name three ways.

a. A piece of paper with _____ rows, each row containing _____ roses.

b. A piece of paper with _____ rows, each row containing _____ roses.

c. A piece of paper with _____ rows, each row containing _____ roses.

On this small piece of wrapping paper, there are four rows, each containing $5\frac{1}{2}$ dogs, which gives a total of 22 dogs.

4. Identify three different ways to cut a piece of this type of wrapping paper so that it would contain 800 dogs.

a. A piece of paper with _____ rows, each row containing _____ dogs.

b. A piece of paper with _____ rows, each row containing _____ dogs.

c. A piece of paper with _____ rows, each row containing _____ dogs.

5. Fill in the blanks to make each multiplication problem correct.

a. _____ × _____ = 600

b. _____ × _____ = 90

c. _____ × _____ = 160

d. _____ × _____ = 720

e. _____ × _____ = 1,200

f. _____ × _____ = 840

Name ______________________ Date ____________

Tiles (Page 1 of 2)

A new school building is being built with a different tile design for each hallway.

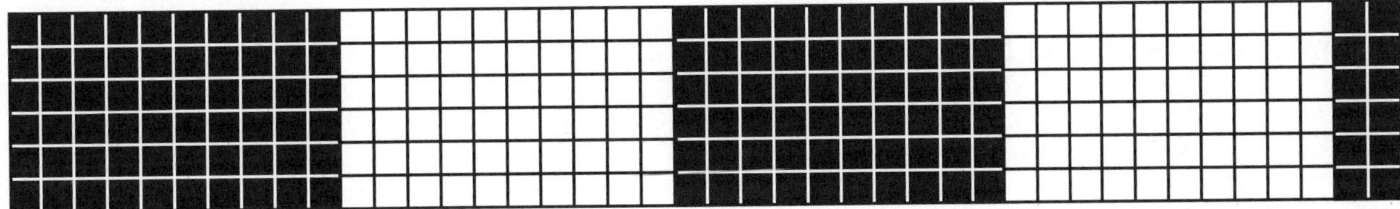

1. Above is the tile design for one hallway on the first floor. It contains six rows of 42 tiles each. How many individual tiles will this hallway have?

2. A hallway on the second floor is the same width, but longer. The tile design for this hallway contains six rows of 77 tiles each. How many individual tiles will this hallway have?

3. One of the hallways on the third floor is just as long as the one on the first floor, but it is one and a half times as wide. It will have nine rows of 42 tiles each. How many individual tiles is this?

4. Compare your answer to problem **3** with your answer to problem **1.** The number of tiles needed for the third-floor hallway should be one and a half times the number needed for the first-floor hallway. Is it? Explain why it should be.

Name ____________________ Date ____________

Tiles (Page 2 of 2)

5. A hallway on the fourth floor is twice as wide and twice as long as the first-floor hallway. It will have 12 rows of 84 tiles each. How many individual tiles is this?

6. Compare your answer to problem **5** with your answer to problem **1.**

7. How does a change in the width or a change in the length of a hallway affect the total number of tiles?

8. For a hallway six tiles wide, 204 tiles are needed. How long do you think the hallway is? (Hint: Making a sketch may help.)

9. There is a hallway on the fourth floor that will have 184 tiles. That hallway is 23 tiles long. How wide is it?

Name ______________________ Date ____________

Section C. Number Sense

Blots (Page 1 of 2)

Alphonse and Sarah decided to do their homework together. When they took out the one copy of the activity sheet they had, they discovered that it had gotten wet, and the ink had blotted some of the numbers. One of the problems looked like this:

6 x 4■ =

At first they thought they could not do the problem. Then, Alphonse wrote, "The answer is more than 240 and less than 294." Sarah read his answer and suggested that the answer could also include the numbers 240 and 294.

1. Who is correct? Explain why.

2. Here are more of the problems on the activity sheet. The second number in each problem is a two-digit number. Find the ranges for the answers to these problems.

a. 8 x 3■ =

b. 6 x 5■ =

c. 5 x 7■ =

d. 5 x 2■ =

e. 7 x 8■ =

f. 3 x 3■ =

g. 4 x 2■ =

h. 6 x 4■ =

i. 3 x 9■ =

j. 9 x 3■ =

Name ____________________ Date ____________

Blots (Page 2 of 2)

3. In the following set of problems, the second number always contains three digits. Find the ranges for the answers to these problems.

a. 3 x 3■■ =

b. 2 x 25■ =

c. 6 x 4■2 =

d. 3 x 62■ =

e. 9 x 2■0 =

f. 6 x 54■ =

g. 4 x 6■9 =

h. 3 x ■70 =

i. 8 x 53■ =

j. 7 x 24■ =

Name ______________________ Date ______________

Timber (Page 1 of 1)

At Chucky's Lumberyard, lumber is sold in these lengths: 120 centimeters, 160 centimeters, 200 centimeters, 240 centimeters, and 280 centimeters.

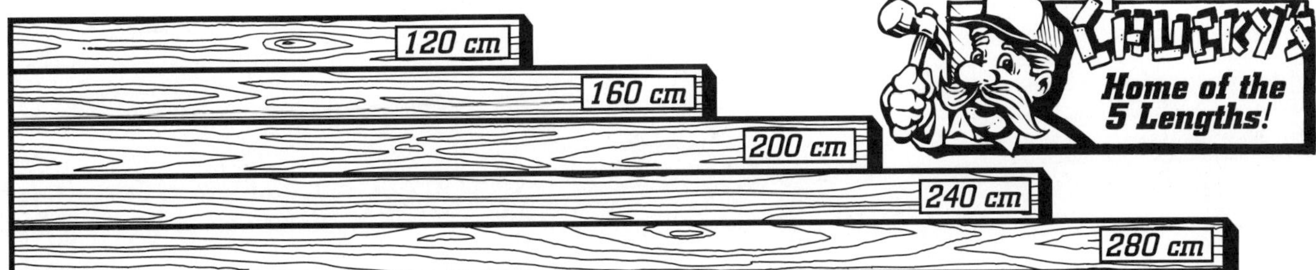

For the following problems, find the shortest piece of wood each person could buy and cut up in order to have the needed lengths.

1. Peter is making a bookcase; he needs eight pieces of wood, each 34 centimeters long.

2. Teresa needs five pieces of wood, each 39 centimeters long.

3. Hua needs four pieces of wood, each 68 centimeters long.

4. Bobby needs three pieces of wood, each 82 centimeters long.

5. Mika needs six pieces of wood, each 41 centimeters long.

6. Avi needs four pieces of wood, each 63 centimeters long.

7. Margaret needs six pieces of wood, each 38 centimeters long.

8. Dieter needs five pieces of wood, each 47 centimeters long.

9. Uma needs 12 pieces of wood, each 21 centimeters long.

10. Jennifer needs nine pieces of wood, each 28 centimeters long.

BRITANNICA Mathematics in Context

Section D. Calculator Activities

Name ______________________ Date ______________

Section D. Calculator Activities

Do You Really Need a Calculator? (Page 1 of 1)

For each of the following story problems, decide whether or not you need to use a calculator. Then solve the problem. If you used a calculator to solve the problem, write "calculator" after your answer.

1. Ms. Weafry found an envelope containing twenty $10 bills. What is the total amount of money Ms. Weafry found?

2. On a shopping trip for his parents, Darnell purchased items worth $4.32, $3.76, $2.58, and $3.84. What was the total cost of the four items?

3. When entering a Mosque (a Muslim religious building), people take off their shoes. When 25 people have entered an empty Mosque, how many individual shoes remain at the entrance?

4. At a supermarket checkout, a woman has three items in her cart worth $3.75, $6.92, and $3.83. In her wallet, she finds she has only a $10 bill and a $5 bill. Does she have enough money to pay for the items?

5. Jacqueline earns $4.60 each time she washes her mother's car. How much money will she earn if she washes the car five times?

6. The school auditorium has 990 seats. There are 45 seats in each row. How many rows are there in the auditorium?

7. A can of baked beans costs 98¢. How much would you have to pay for 25 cans of baked beans?

8. The book John is reading contains 416 pages. Yesterday he read 158 pages, and today he read 117 more pages. How many pages must he read tomorrow to finish the book?

Name ______________________ Date ______________

Calculator Calculating (Page 1 of 1)

1. Use your calculator to determine the missing numbers in the problems below.

   a. 12,345 − __________ = 4,321

   b. 12,345 + __________ = 54,321

   c. __________ − 12,345 = 66,666

   d. 1,102,914 ÷ __________ = 213

   e. __________ × 345 = 26,910

   f. __________ ÷ 362 = 17

   g. 45,120 ÷ 23 = __________ remainder 17

   h. 12,431 ÷ 13 = 956 remainder __________

2. Use your calculator to determine the missing numbers. Only **whole** numbers greater than 1 may be used.

   a. ______ × ______ = 403

   b. ______ × ______ = 169

   c. ______ × ______ = 161

   d. ______ × ______ = 1,247

3. Use your calculator to find one digit for each blank that makes each statement correct.

   a. ___,438 × 43___ = 1,048,340

   b. ___5___ × 72___ = 257,744

   c. 1___9 × 2___ = 3,___25

Name ______________________________ Date ______________

How Old? (Page 1 of 1)

Suppose it is your birthday. People will probably ask how old you are, and your answer will be a number of years. Could you answer in terms of days, hours, and/or seconds? Use a calculator for the following problems. If the calculator's screen or display is too small for all of the digits, devise another way to answer the problems.

1. How old will you be in years on your next birthday?

2. How old will you be in days on your next birthday?

3. How old will you be in hours on your next birthday?

4. How old will you be in seconds on your next birthday?

5. How many seconds old will you be on your 60th birthday?

6. How many years old will the United States be next Independence Day? (Hint: The United States became independent in 1776.)

7. How many seconds old will the United States be next Independence Day?

Name ______________________ Date ______________

In a Shorter Way (Page 1 of 1)

To solve each of the following problems, first think about how you can most efficiently use your calculator to get the answer. This means finding a shorter way to calculate the problem than entering each number and symbol into your calculator as given. Write down your method for solving each problem and then find the answer.

1. 324 + 268 + 268 + 268 + 268 + 268 + 268 + 268 + 268 =

2. 123 + 89 + 123 + 89 + 123 + 89 + 123 + 89 + 123 + 89 + 123 + 89 =

3. 3 × 1,342 + 3 × 1,342 + 4 × 1,342 + 3 × 1,342 + 7 × 1,342 + 3 × 1,342 =

4. 17 × 351 + 31 × 351 + 9 × 351 + 36 × 351 + 12 × 351 + 9 × 351 =

5. 1 + 2 + 3 + 4 + 5 + 6 + 7 + 8 + 9 + 10 +

11 + 12 + 13 + 14 + 15 + 16 + 17 + 18 + 19 + 20 +

21 + 22 + 23 + 24 + 25 + 26 + 27 + 28 + 29 + 30 +

31 + 32 + 33 + 34 + 35 + 36 + 37 + 38 + 39 + 40 +

41 + 42 + 43 + 44 + 45 + 46 + 47 + 48 + 49 + 50 +

51 + 52 + 53 + 54 + 55 + 56 + 57 + 58 + 59 + 60 +

61 + 62 + 63 + 64 + 65 + 66 + 67 + 68 + 69 + 70 +

71 + 72 + 73 + 74 + 75 + 76 + 77 + 78 + 79 + 80 +

81 + 82 + 83 + 84 + 85 + 86 + 87 + 88 + 89 + 90 +

91 + 92 + 93 + 94 + 95 + 96 + 97 + 98 + 99 + 100 =

6. Try to find another way to calculate the answer to problem **5.**

Name ______________________ Date ______________

Calculator Puzzles (Page 1 of 1)

1. Enter the number 342,477 into your calculator.
   a. Using your calculator, wipe out the 3 in 342,477 so that the result is 42,477. Explain how you did this.
   b. Using your calculator, change all the 4s in the number 342,477 into 5s. Explain how you did this.
   c. Using your calculator, wipe out the 7s in 342,477 so that the result is 3,424. Explain how you did this.

2. Enter any four-digit number into your calculator. Using the operations of your calculator, wipe out each of the digits one by one. Explain how you did this.

3. To solve each of the following puzzles, you can use only four buttons on your calculator:

You may use each button as often as you like, and you may use the buttons in any order.

   a. Using the four buttons, make the number 99 appear in your calculator display in as few steps as possible. Explain how you did this.
   b. Clear the calculator. Now make the number 9,999 appear in as few steps as possible. Explain how you did this.
   c. Clear the calculator. This time make the number 123,456 appear in as few steps as possible. Explain how you did this.
   d. Are there any numbers you cannot create if you are restricted to these four buttons? Explain.

Name ______________________ Date ______________

Looking at It Upside Down (Page 1 of 2)

Mark entered the number 808 into his calculator and found that when he turned his calculator 180°, or upside down, the number in the display looked the same.

1. a. Find another three-digit number that looks the same upside down.

b. Find a four-digit number that looks the same upside down.

c. Find a five-digit number that will look the same upside down.

2. Enter 696,969 into your calculator. Turn your calculator 180°.

a. What appears in the display now?

b. Using the same two digits, 6 and 9, enter a six-digit number that will become a smaller number when you turn the calculator 180°.

c. Using only the digits 6 and 9, enter a six-digit number that becomes the largest possible number that fits in the calculator's display after turning the calculator upside down.

Name ______________________ Date ______________

Looking at It Upside Down (Page 2 of 2)

3. Enter 77,345 into your calculator. Turn your calculator 180°. What word appears in the display?

4. The answer to each of the following calculations, when read upside down, is a word. What is each word?

a. $23 \times 2 \times 2 \times 2 \times 2 - 23$

b. $266 \times 4 \times 5 - 3$

5. There are many calculation problems whose answers, when read upside down, are words.

a. Find a calculation problem whose answer will result in the word ShIELDS when you turn the calculator 180°.

b. Explain why it is not possible to get the word ShIELD using this process.

c. Create a set of calculations whose answer is a word and exchange your calculations with a classmate.

Name ______________________ Date ____________

1, 2, 3, 4, 5 (Page 1 of 1)

1. In the following exercises, use each of the digits 1, 2, 3, 4, and 5 only once to make a true statement. Here are three examples:

12 × 345 = 4,140 315 × 24 = 7,560 3 × 4,521 = 13,563

Now it is your turn:

a. ___ ___ ___ × ___ ___ = 5,220

b. ___ ___ ___ × ___ ___ = 10,793

c. ___ ___ ___ × ___ ___ = 10,325

2. In the following exercises, use each of the digits 5, 6, 7, 8, and 9 only once to make a true statement.

a. ___ ___ ___ × ___ ___ = 60,075

b. ___ ___ ___ × ___ ___ = 74,670

c. ___ ___ ___ × ___ ___ = 56,842

Name ______________________ Date ______________

The Goal Game (Page 1 of 1)

To solve a problem, Simone used each of the following calculator buttons once:

The answer to her calculation was the number 636.

1. Can you figure out in what order Simone pressed the buttons? There is only one possibility that produces an answer of 636.

2. Play the following game in pairs or small groups.

Game Rules: You may use only the following buttons:

1 2 3 4 5 6 7 8 9 0 + x =

Step 1: One student, the *goalkeeper,* performs a calculation on a calculator with exactly five button strokes.

Step 2: The goalkeeper names the buttons he or she used and how often each was used, but not the order in which they were used. The goalkeeper then shows the other students the result displayed on the calculator: this is called the *goal number.*

Step 3: Using the information provided by the goalkeeper, the other players have two minutes to think about what the calculation might be and then—at the end of the two minutes—one try on the calculator.

Scoring: Each player who comes up with the correct calculation has scored a goal and is awarded one point.

Note: Players take turns being the goalkeeper. The player with the most points at the end of a set number of rounds wins the game.

BRITANNICA Mathematics in Context

Section E. Number Knowledge

Name ______________________ Date ______________

Lottery Tickets (Page 1 of 2)

This year, Johnson School is celebrating its 80th anniversary. The celebration includes a large party to which all of the students' parents have been invited. During the party, students will sell lottery tickets to raise money for the school library.

Each student has been given tickets to sell, and each ticket is numbered. Pictured below is the packet of tickets Sandra will try to sell at the party.

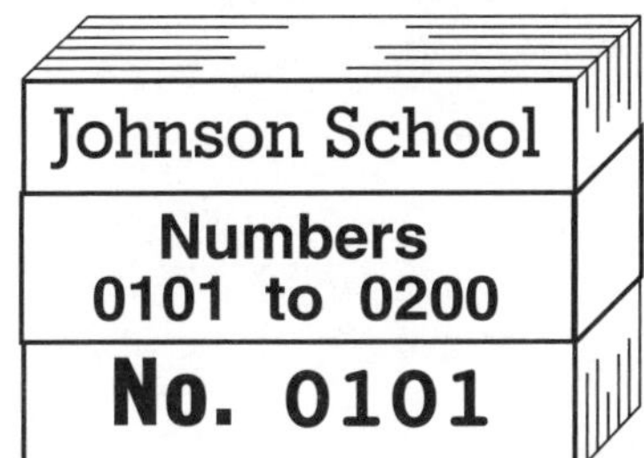

1. List a few of Sandra's ticket numbers.

Gill and Mary are also selling lottery tickets. Gill's tickets are numbered 0401 to 0500. Mary's tickets are numbered 1201 to 1300.

2. List a few of Gill's ticket numbers.

3. List a few of Mary's ticket numbers.

4. Do Sandra, Gill, and Mary have the same number of tickets to sell? Explain.

Chantrea is also selling lottery tickets. After selling some tickets, the number of the ticket on top of her packet is 3414.

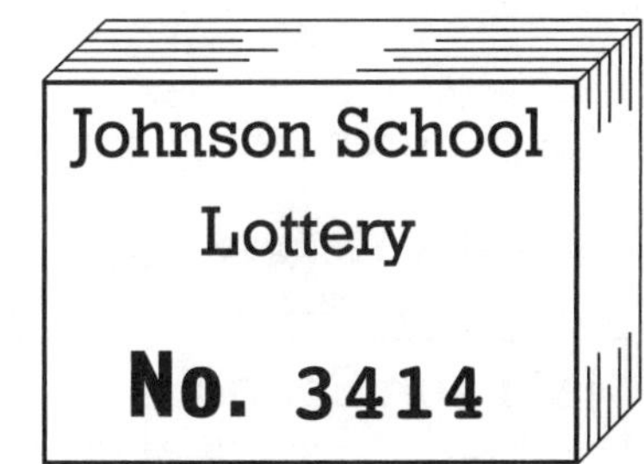

5. What ticket numbers were in Chantrea's original packet? How do you know?

6. How many tickets has Chantrea sold?

Name ______________________ Date ______________

Lottery Tickets (Page 2 of 2)

Mr. Han is one of the parents attending the party.
He buys several tickets, which are pictured on the right.

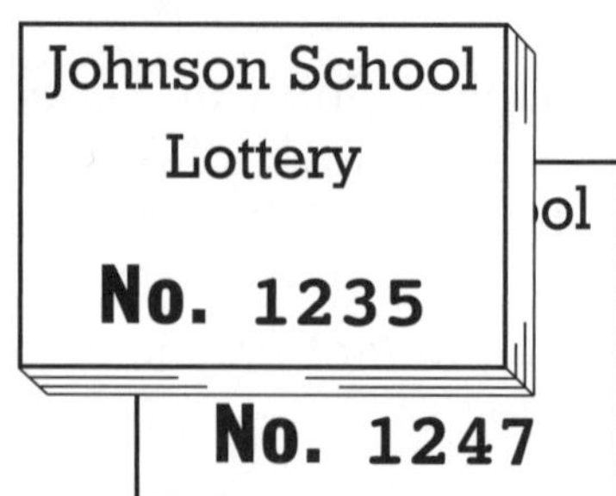

7. If the number of his first ticket is 1235, and the number of his last ticket is 1247, how many tickets did Mr. Han buy?

8. From which student did Mr. Han buy his tickets? How do you know?

When the party ends, Sandra has some tickets left. She has sold up to ticket number 0178.

9. How many tickets does Sandra have left?

The pictures below show Gill's and Mary's leftover tickets.

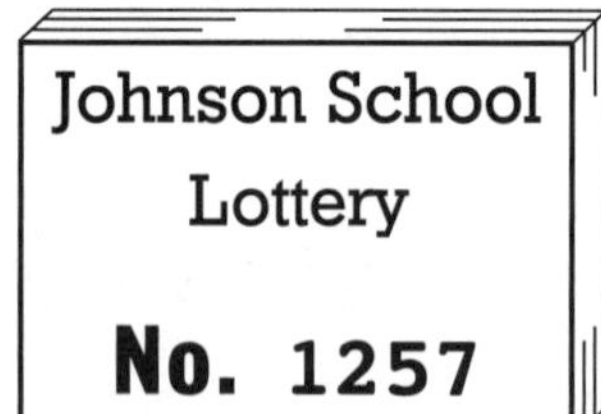

10. How many tickets does each have left?

Name ______________________ Date ______________

Ratio Tables for Multiplying (Page 1 of 1)

1. Ms. Morningstar pays $461 per month to rent her apartment. She wants to know what her rent will be for a full year. Use the following ratio table to find the answer.

Months	1	10	2		
Rent Paid	$461				

2. Enrico lives in San Diego, California, and Loretta lives in Barcelona, Spain. Enrico called Loretta, and they talked on the phone for 23 minutes. If it costs 83¢ a minute for Enrico to call Barcelona, how much did the phone call to Loretta cost? Use the following ratio table to find your answer.

Minutes	1				
Price	$0.83				

3. Rashad rented a safe-deposit box at a bank for 65¢ a day. If he pays the rent for an entire year, what amount will he have to pay? Use the following ratio table to find your answer.

Days	1				
Price	$0.65				

4. Make up your own story that requires the calculation 43 × 204. Find the answer to your problem using a ratio table.

Name ______________________ Date ______________

Ratio Tables for Dividing (Page 1 of 1)

Solve each of the following problems using the given ratio table.

1. A school ordered 17,051 notebooks. The notebooks were delivered in cartons containing 17 notebooks each. Assuming that all of the cartons were full, how many cartons did the school receive?

Cartons	1	1,000			
Notebooks	17				

2. A bus company must take 4,853 people from their hotels to a convention center at one time. Each bus can carry up to 23 passengers. How many buses are needed?

Buses	1	100			
People	23				

3. An orchestra won a prize of $8,339. The 31 members of the orchestra decide to share the prize equally. How much will each member receive?

Each Member	$1				
Total	$31	$3,100			

4. Make up your own story that requires the calculation 40,336 ÷ 16. Find the answer to your problem using a ratio table.

Name ______________________ Date ______________

Easy Numbers (Page 1 of 1)

1. The attendance at a baseball game was 30,410 people. Each person paid $10 for a ticket. Use the following ratio table to find the total ticket revenue for the game.

People				
Revenue				

2. Ms. Peluy withdrew $24,000 from her bank account in $100 bills. Use the following ratio table to find out how many $100 bills Ms. Peluy received from the bank.

Bills				
Money				

3. There are 5,000 students who take the bus to school. Each bus can carry 50 students. How many buses are required? Use the following ratio table to solve this problem.

Buses				
Students				

4. Suppose you have a candle that is 125 centimeters tall. It takes 10 days to completely burn this candle. How much of the candle burns away each day? Use the following ratio table to solve this problem.

Days				
Lengths				

5. Solving each of the problems above should have taken very few steps. What makes these problems easy to solve?

6. Answer the following problems without using a ratio table or a calculator:

   a. 15,900 ÷ 100 =

   b. 3,000 ÷ 30 =

   c. 120 ÷ 100 =

   d. 450 × 1,000 =

Name ________________________ Date ____________

Serial Numbers (Page 1 of 1)

The Green Air factory makes refrigerators. Each refrigerator is given a serial number. Here are the serial numbers of the first three refrigerators made last Tuesday:

SR-341-05-0193
SR-341-05-0194
SR-341-05-0195

1. List the serial numbers of the next three refrigerators made at the Green Air factory.

Below are the serial numbers of the last three refrigerators made last Tuesday.

SR-341-05-3601
SR-341-05-3602
SR-341-05-3603

2. How many refrigerators were made last Tuesday?

Last Wednesday, the first refrigerator made was given the serial number **SR-341-05-3604,** and the last refrigerator had the serial number **SR-341-05-7702.**

3. How many refrigerators were made last Wednesday?

The last refrigerator made on Thursday had the serial number **SR-341-05-9871,** and the last refrigerator made on Friday had the serial number **SR-341-06-3004.**

4. How many refrigerators were made on Thursday? on Friday?

Name ______________________ Date ______________

Five Little Stories (Page 1 of 1)

1. A school needs to purchase 513 new computers. If each computer costs $3,470, what is the total cost of these new computers? Use your calculator to solve this problem.

To solve each of the following problems, do not use your calculator. Instead, use your answer to the first problem as a starting point.

2. A ticket to the circus costs $5.13. If the attendance at the circus was 347 people, what was the total ticket revenue?

3. Every year, a ferryboat sails 3,470 times from the mainland to the island of Olku. The maximum number of passengers allowed on the ferry is 5,130. What is the maximum number of people that the boat can carry to the island each year?

4. Pierre pays $34.70 for one kilogram of specialty mushrooms for his restaurant. What would he pay for 5.13 kilograms of mushrooms?

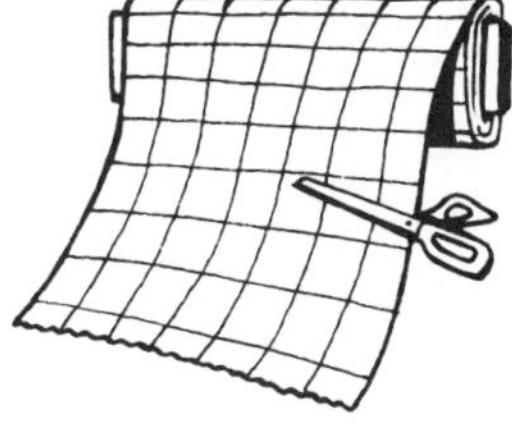

5. Mr. Flores is making some of the costumes for the school play. He needs 51.3 yards of fabric, and the fabric costs $3.47 a yard. What is the total cost for the fabric?

Name ____________________ Date ____________________

Close Enough (Page 1 of 1)

For each of the following problems, draw a circle around the number closest to the correct answer. You should not use your calculator or make precise calculations to find the answer. Explain how you chose each answer.

1. 101 × 11

800
900
1,000
1,100
1,200

2. 391 × 391

10
100
1,000
10,000
100,000
1,000,000

3. 111 × 909

800
9,000
10,000
110,000
1,200,000

4. 91 × 19 × 19

500
5,000
50,000
500,000
5,000,000

5. 30 × 41 × 52

100
1,000
10,000
100,000
1,000,000

6. 1,234 × 5,678

1,000
10,000
100,000
1,000,000
10,000,000

Name ______________________ Date ____________

Certificates (Page 1 of 2)

Climbers who reach the top of Mt. Erghoog are awarded a certificate when they return to the park station. Because Mt. Erghoog is a very popular mountain to climb, the park station keeps many certificates on hand. These certificates are shipped to the park station in large boxes. Each large box contains 10 small boxes. Inside each small box are 10 envelopes. Inside each envelope are 10 certificates.

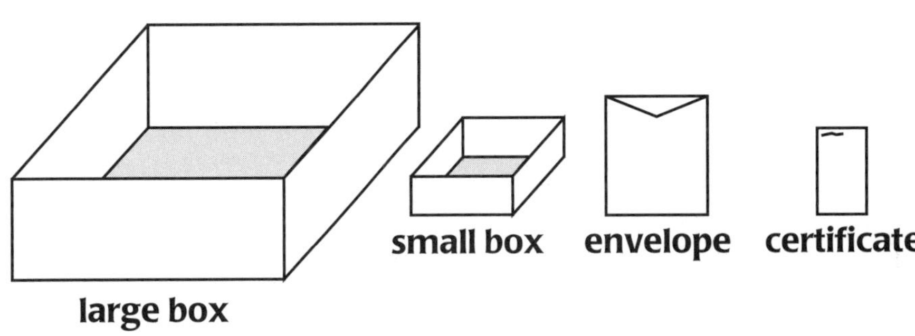

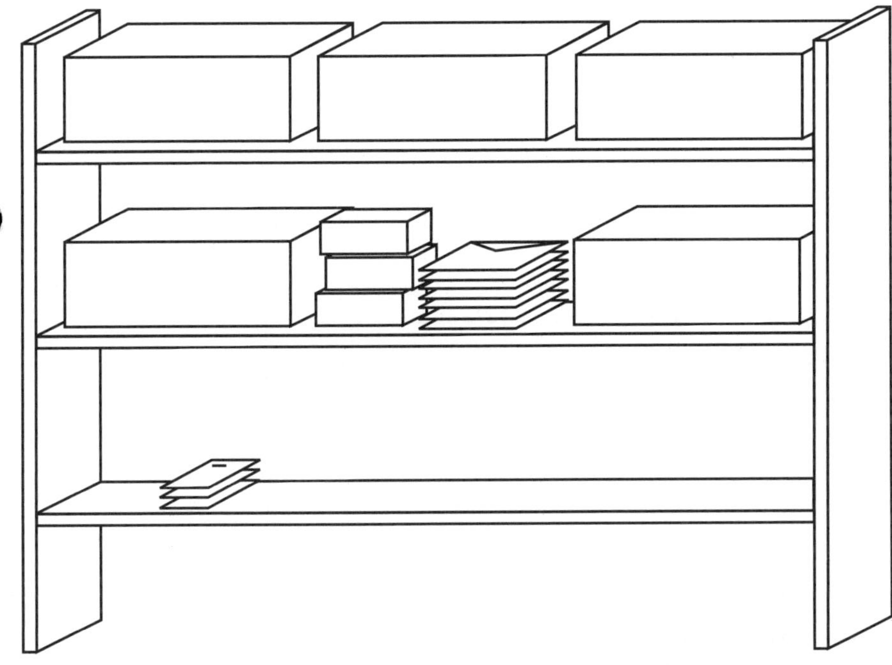

An inventory of the certificates is taken at different times during the year. On May 1, the inventory consisted of what you can see in the picture on the left. There are three loose certificates on the bottom shelf, and all the boxes and envelopes are full.

The inventory is recorded in a table similar to the one below.

1. Complete the table.

Date	Large Boxes	Small Boxes	Envelopes	Certificates
May 1				

2. How many certificates were in stock on May 1?

Name ______________________ Date ______________

Certificates (Page 2 of 2)

3. On July 1, inventory was taken again. There were 29 large boxes, 16 small boxes, 8 envelopes, and 29 loose certificates. According to the books, there should have been 30,709 certificates in stock. Was the actual number of certificates in agreement with the books? Explain.

4. On September 1, there should have been 2,015 certificates in stock, according to the books. Use the table below to show different ways in which this inventory could have been stored on the shelves.

Large Boxes	Small Boxes	Envelopes	Certificates

Name ______________________________ Date ______________

Section E. Number Knowledge

At the Cash Register (Page 1 of 1)

For the following problems, think of yourself as a cashier at a store. Customers are paying you for items, and you are giving them change. When you make change, you should hand back the fewest bills and coins possible.

1. In the following table, tally the coins and/or bills you would return to the customer for each transaction. The first row has been done for you.

Price	Payment	1¢	5¢	10¢	25¢	$1	$5	$10	$20	$50
$13.49	$20.00	/			//	/	/			
$67.88	$100.00									
$198.21	$200.01									
$23.02	$100.00									
$93.55	$100.00									

2. For the following table, first choose how much the customer pays and then decide what money would be returned.

Price	Payment	1¢	5¢	10¢	25¢	$1	$5	$10	$20	$50
$5.98										
$16.23										
$59.80										
$42.95										
$5.08										

3. For the following table, choose both the price and the amount the customer pays. After filling in the first two columns, fill in the rest of the table for the amounts of money returned.

Price	Payment	1¢	5¢	10¢	25¢	$1	$5	$10	$20	$50
$										
$										
$										
$										
$										

Section F. Fractions

Name ______________________ Date ____________

Shoveling Sidewalks (Page 1 of 1)

When it snows, Raúl and his friends help their neighbors by shoveling their sidewalks. Each person who shows up to help shovels an equal area. The pictures below show how many people are helping and how many sidewalks need to be shoveled on different days. Color the sidewalks for the following problems and use fractions to show the areas each helper will have to shovel.

Sidewalks | **People Helping**

1. **Each helper shovels:**

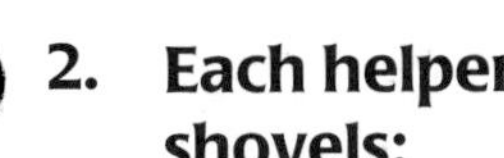

2. **Each helper shovels:**

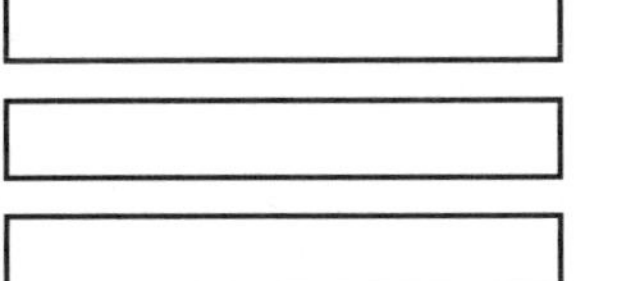

---

3. **Each helper shovels:**

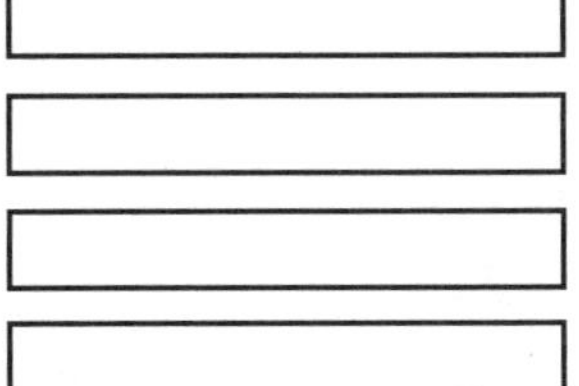

---

4. **Each helper shovels:**

Name ______________________ Date ______________

Section F. Fractions

Chocolate Bars (Page 1 of 1)

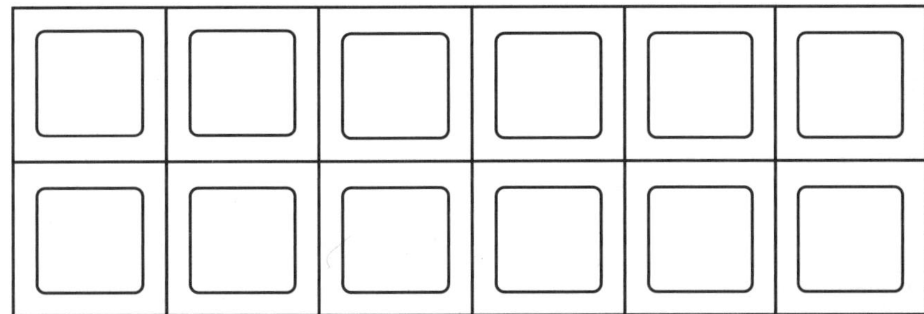

1. Mike, Liz, and Jordan are sharing a chocolate bar. The chocolate bar has a total of 12 sections, of which each person will get four. Write a fraction to describe how much of the whole bar each person will get.

2. For each example below, write a fraction to represent the part of the bar that is shaded.

a.  

b. 

c. 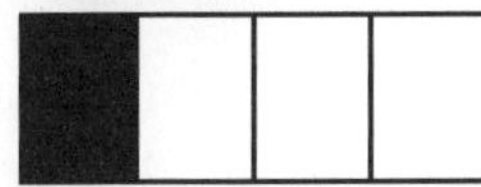

d. 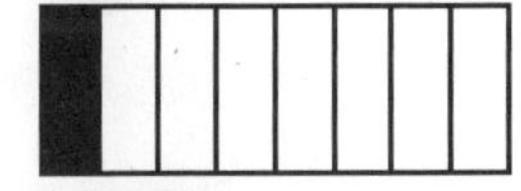

e. 

f. 

g. 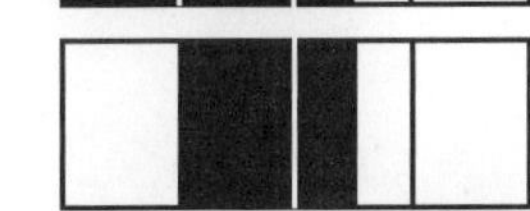

h.

i. 

j. 

k. 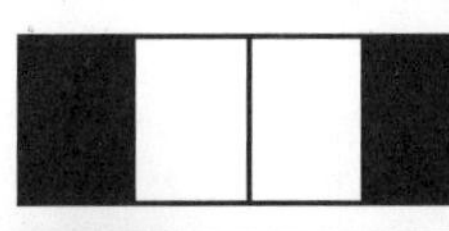

l. 

m. 

n. 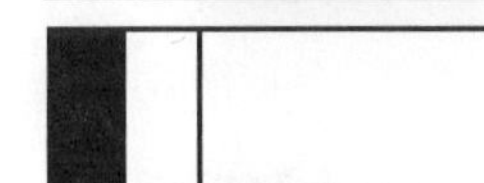

o. 

p. 

q. 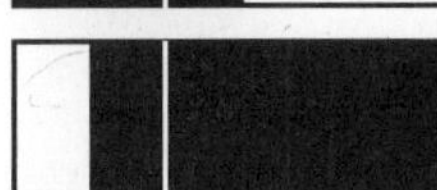

r. 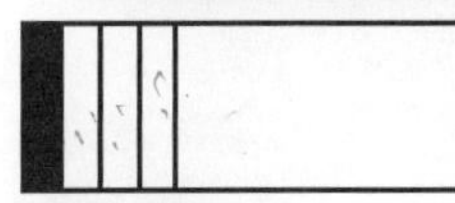

s.

t. 

Name ______________________ Date ______________

Measuring Cups (Page 1 of 1)

On the right is a picture of the measuring cup that Jeremy uses when he is cooking. It is not very precise because it only has markings that indicate 1 cup and 2 cups. If a recipe asks for $\frac{1}{3}$ cup or $1\frac{1}{4}$ cup, Jeremy has to estimate.

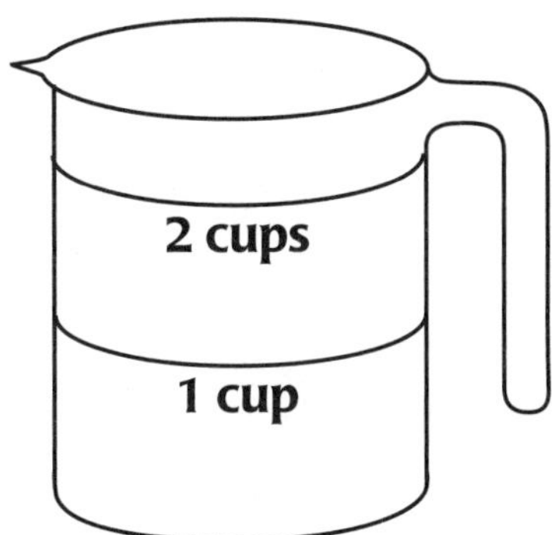

Shade each measuring cup below to show about where each fraction of a cup or combination of fractions would be.

1. $\frac{1}{4}$ cup

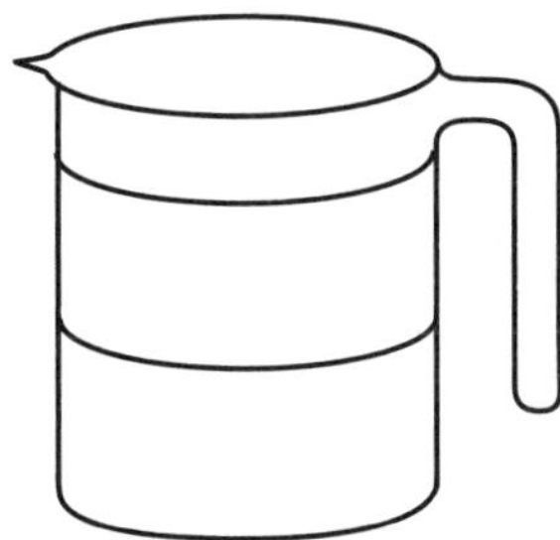

2. $\frac{3}{8}$ cup

3. $\frac{1}{3}$ cup

4. $\frac{2}{3}$ cup

5. $\frac{3}{4}$ cup

6. $\frac{5}{8}$ cup

7. $\frac{3}{3}$ cup

8. $\frac{3}{4}$ cup + $\frac{1}{2}$ cup

9. $\frac{1}{2}$ cup + $\frac{2}{3}$ cup

10. $\frac{3}{4}$ cup + $\frac{3}{8}$ cup

11. $\frac{1}{4}$ cup + $\frac{3}{4}$ cup

12. $\frac{2}{3}$ cup + $\frac{2}{3}$ cup

East of Yaksee Cave (Page 1 of 2)

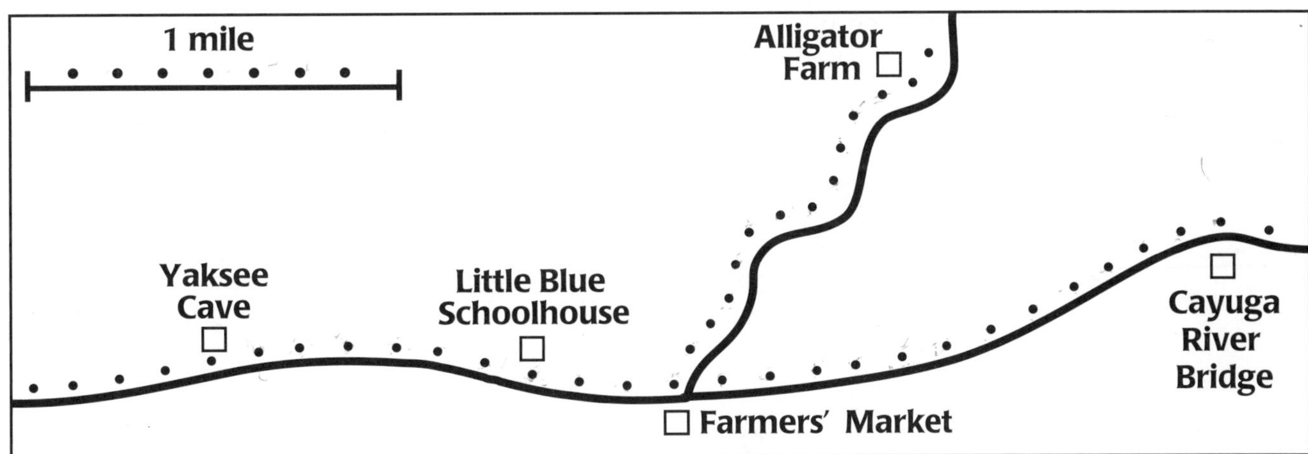

1. Use the above map to enter the distances in the table below. The distance from each place to the Little Blue Schoolhouse is already entered. Whenever possible, express the fractional distances as halves or quarters instead of eighths of a mile.

	Little Blue Schoolhouse	Yaksee Cave	Farmers' Market	Alligator Farm	Cayuga River Bridge
Little Blue Schoolhouse	0	$\frac{7}{8}$ mile	$\frac{3}{8}$ mile	$1\frac{7}{8}$ miles	2 miles
Yaksee Cave					
Farmers' Market					
Alligator Farm					
Cayuga River Bridge					

Name ______________________ Date ______________

East of Yaksee Cave (Page 2 of 2)

2. a. Make up and answer three addition problems using some of the fractions below. You may use the scale on the map on page 119 to help you add the fractions.

$\frac{7}{8}, \frac{3}{8}, \frac{1}{4}, 1\frac{1}{2}, \frac{1}{8}, \frac{5}{8}, \frac{3}{4}, \frac{1}{2}$

b. Write three more addition problems. This time, be sure to combine eighths, quarters, and/or halves.

Example: $1\frac{1}{2} + \frac{3}{8} = 1\frac{7}{8}$

Name ________________________ Date ____________

Passing by Tom's House (Page 1 of 1)

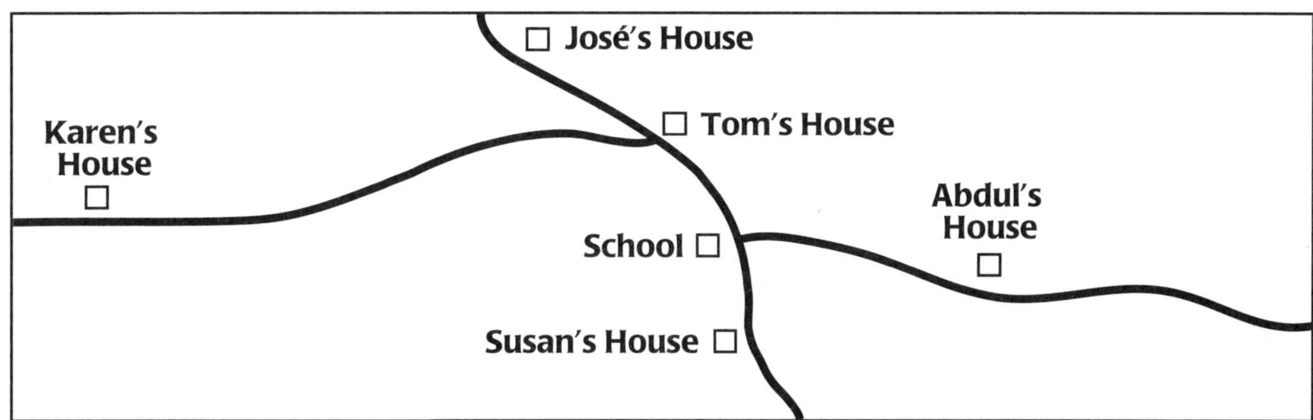

1. This map does not show a scale line; however, the distance from each person's house to the school is given below. Enter the missing distances in the table. Whenever possible, express the distances as halves or quarters instead of eighths of a mile.

	School	Tom	José	Karen	Abdul	Susan
School	0	$\frac{3}{4}$ mile	$1\frac{1}{2}$ miles	3 miles	$1\frac{1}{8}$ miles	$\frac{5}{8}$ mile
Tom						
José						
Karen						
Abdul						
Susan						

2. Write three addition and three subtraction problems involving eighths, quarters, and/or halves.

Example: $1\frac{1}{2} - \frac{3}{4} = \frac{3}{4}$

Name ______________________ Date ____________

Walking Trail (Page 1 of 1)

The Wetlands Preserve is creating a walking trail that will be 12 kilometers long. At different places along the trail, there will be rest areas and areas for wildlife-viewing.

a. Picnic Area
$\frac{1}{2}$ of the way

b. Bee Colony
$\frac{1}{6}$ of the way

c. Restrooms
$\frac{2}{3}$ of the way

d. Dinosaur Footprint
$\frac{1}{3}$ of the way

e. Egret Nesting Area
$\frac{3}{4}$ of the way

f. Wilderness Campground
$\frac{5}{6}$ of the way

g. Butterfly Meadow
$\frac{1}{4}$ of the way

1. Use the line below to represent the trail. Write the letter that represents each place at its location along the trail.

Beginning of Trail —————————————————— **End of Trail**

2. How many kilometers from the beginning of the trail is each place located?

3. To the right is a map of the trail. Write the letter that represents each place at its location along the trail.

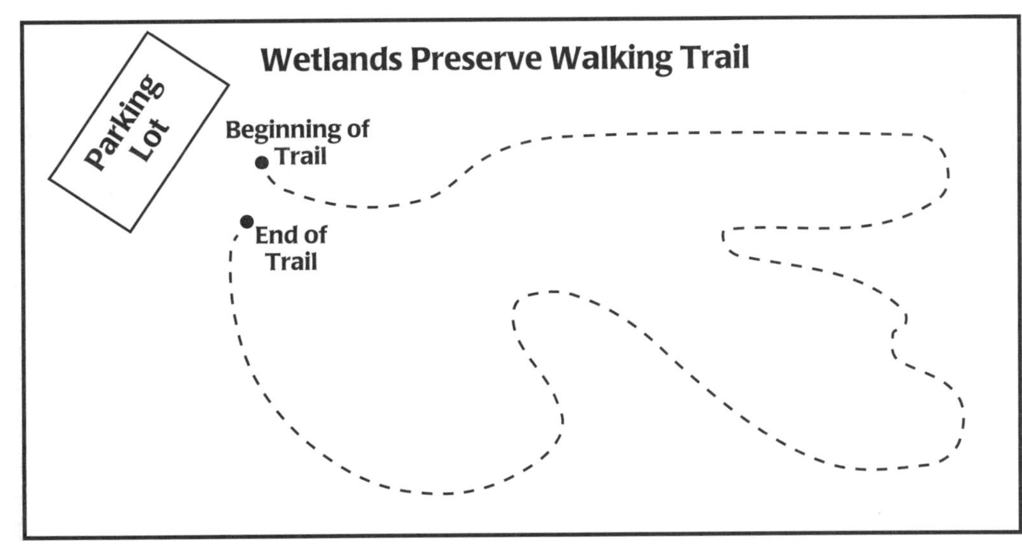

Name ______________________ Date ______________

Driving to Mons (Page 1 of 1)

The town of Mons is near five other cities. The map on the right shows the driving times required to reach these cities from Mons.

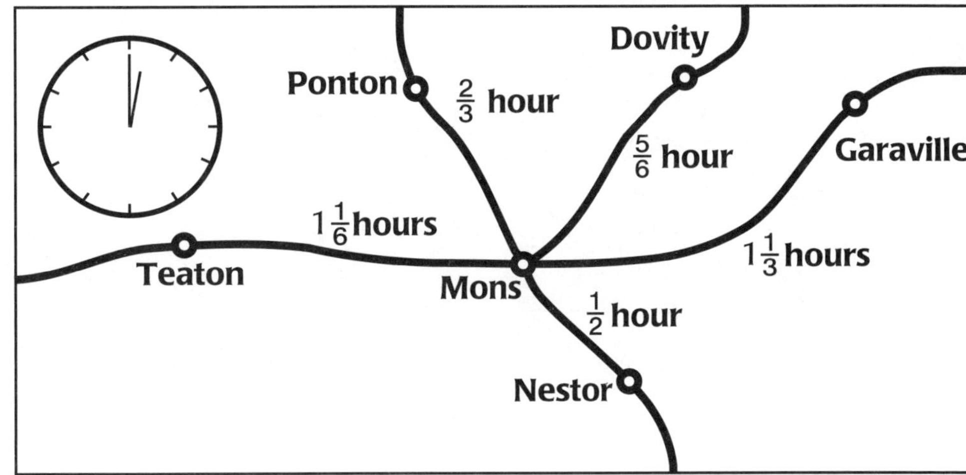

1. Using the above map, fill in the driving times between the pairs of cities in the table on the right. The driving time from Mons to each city has been filled in.

	Mons	Ponton	Dovity	Garaville	Nestor	Teaton
Mons	0	$\frac{2}{3}$ hour	$\frac{5}{6}$ hour	$1\frac{1}{3}$ hours	$\frac{1}{2}$ hour	$1\frac{1}{6}$ hours
Ponton						
Dovity						
Garaville						
Nestor						
Teaton						

2. a. Make up and answer three addition problems using some of the fractions below:

$\frac{1}{3}, \frac{5}{6}, \frac{1}{2}, \frac{1}{6}, 1\frac{1}{6}, 2\frac{1}{3}, \frac{2}{3}$

b. Write three other addition problems that combine sixths, thirds, and/or halves.

Example: $1\frac{1}{3} + \frac{1}{6} = 1\frac{1}{2}$

Name ______________________ Date __________

From Lewisville to Knox (Page 1 of 1)

1. It takes $1\frac{1}{4}$ hours to drive from Peterston to Lewisville, and it takes $\frac{1}{3}$ hour to drive from Peterston to Knox. How long does it take to drive from Lewisville to Knox?

2. How long does it take to drive from Peterston to Knox and back?

3. How long does it take to drive from Peterston to Lewisville and back?

4. Solve the following addition problems:

   a. $2\frac{1}{4}$ hours + $\frac{1}{3}$ hour =

   b. $1\frac{1}{2}$ hours + $\frac{1}{3}$ hour =

   c. $\frac{1}{6}$ hour + $\frac{2}{3}$ hour =

   d. $\frac{1}{6}$ hour + $\frac{1}{4}$ hour =

   e. $2\frac{2}{3}$ hours + $\frac{1}{6}$ hour =

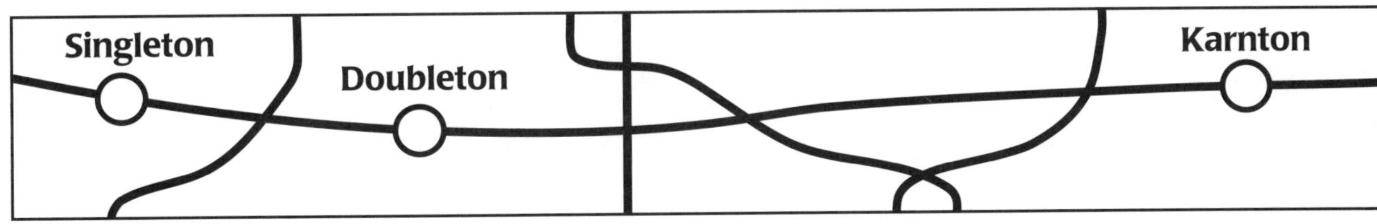

5. Driving from Singleton to Karnton takes $2\frac{1}{2}$ hours. Driving from Singleton to Doubleton takes $\frac{2}{3}$ hour. How long does it take to drive from Doubleton to Karnton?

6. Write and solve five subtraction problems involving halves, quarters, thirds, and/or sixths.

   Example: $1\frac{1}{2} - \frac{2}{3} = \frac{5}{6}$

Sharing Shakes (Page 1 of 1)

Banana Shake

4 servings:

2 bananas
$\frac{1}{3}$ cup lemon juice
$\frac{1}{4}$ cup granulated sugar
$\frac{1}{2}$ quart milk
1 cup vanilla ice cream

Blend until smooth.

Tanisha's favorite summertime drink is a banana shake. She started the following chart so she would know how to mix up enough for different numbers of servings. Fill in the rest of the chart according to the above recipe.

Servings	4	8		6			
Bananas	2		1		6		
Lemon Juice (cups)	$\frac{1}{3}$						$\frac{1}{12}$
Sugar (cups)	$\frac{1}{4}$						
Milk (quarts)	$\frac{1}{2}$					2	
Ice Cream (cups)	1		$\frac{1}{2}$		3		$\frac{1}{4}$

Alonzo's Punch (Page 1 of 1)

Strawberry Punch

4 servings:

$\frac{2}{3}$ cup lemon juice

$\frac{1}{4}$ cup granulated sugar

$1\frac{1}{3}$ cups strawberries, sliced

4 cups ginger ale

16 ice cubes

Stir ingredients together in a punch bowl.

Alonzo likes to make punch for his friends on a hot summer day. He started the following chart so he will know how to make strawberry punch for different numbers of people. Fill in the rest of the chart according to the above recipe.

Servings	8	4	16		20			
Lemon Juice (cups)	$\frac{2}{3}$			2				
Sugar (cups)	$\frac{1}{4}$						$1\frac{1}{4}$	
Strawberries (cups)	$1\frac{1}{3}$					2		
Ginger Ale (cups)	4							18
Ice Cubes	16							

Name ______________________ Date ______________

The Bus Company (Page 1 of 1)

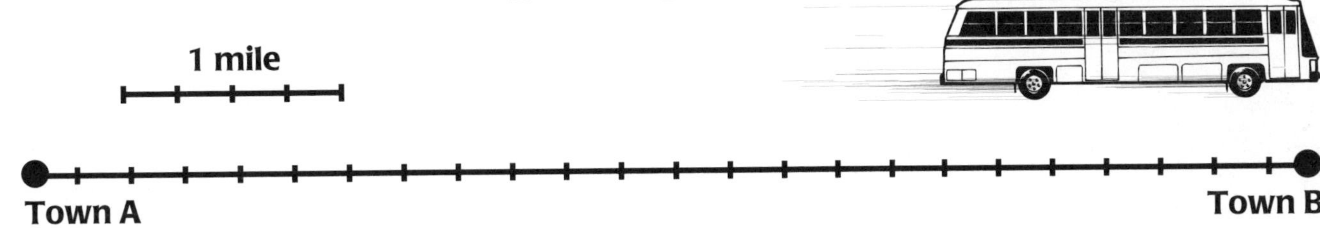

1. A bus company is setting up a route between two towns. The route is 6 miles long, and the company wants to have stops every $\frac{3}{4}$ of a mile. How many stops will there be? Mark the bus stops on the above map.

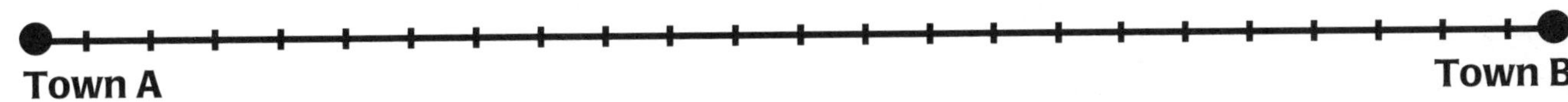

2. How many stops would there be on the same route if the bus stopped every $\frac{3}{8}$ of a mile? Mark the bus stops on the second map, above.

3. Fill in the multiplication table on the right. Whenever possible, simplify your answers. The first column has been done for you.

	$\frac{3}{4}$	$\frac{3}{8}$	$\frac{1}{3}$	$\frac{5}{12}$	$\frac{3}{10}$	$\frac{5}{8}$
1 x	$\frac{3}{4}$					
2 x	$1\frac{1}{2}$					
3 x	$2\frac{1}{4}$					
4 x	3					
5 x	$3\frac{3}{4}$					
6 x	$4\frac{1}{2}$					
7 x	$5\frac{1}{4}$					
8 x	6					
9 x	$6\frac{3}{4}$					
10 x	$7\frac{1}{2}$					

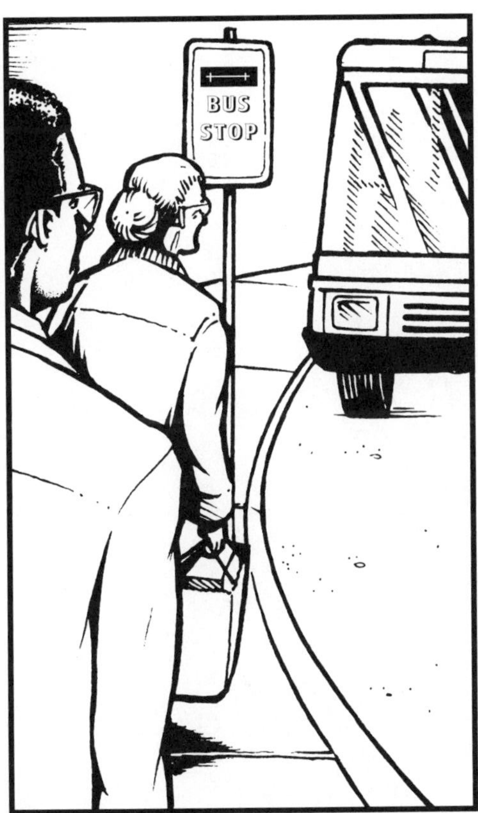

BRITANNICA
Mathematics in Context

Section G.
Decimals

Name ______________________ Date ______________

Gas Tanks (Page 1 of 1)

1. Andrea works at an airport. Her job is to refuel the airplanes. Pictured below are the gasoline gauges from different-sized planes. For each of the gauges, estimate the number of liters of gas the plane has available and what fraction of the plane's capacity this is.

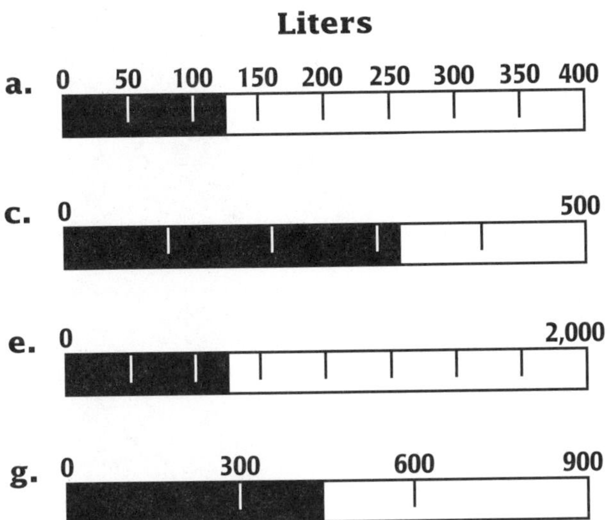

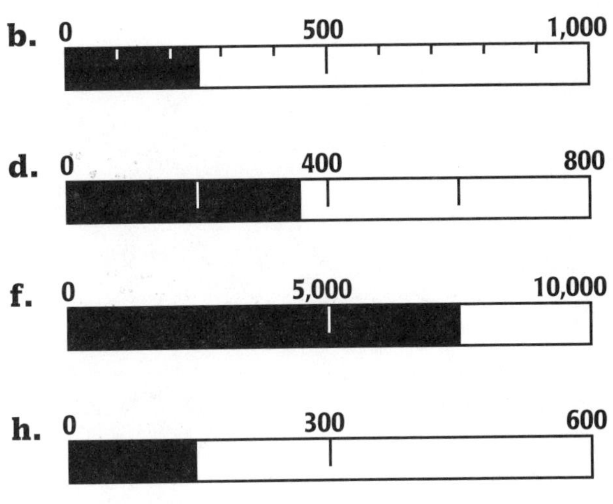

2. An airplane has a 10,000-liter fuel capacity.

   a. If the tanks are $\frac{1}{2}$ full, how many liters of gas are available?

   b. What if the tanks are $\frac{1}{4}$ full?

   c. What if the tanks are $\frac{1}{8}$ full?

3. Another airplane has a 1,200-liter fuel capacity.

   a. If the tanks are $\frac{1}{4}$ full, how many liters of gas are available?

   b. What if the tanks are $\frac{1}{6}$ full?

   c. What if the tanks are $\frac{3}{4}$ full?

4. Another airplane has an 8,000-liter fuel capacity.

   a. If the tanks are $\frac{3}{4}$ full, how many liters of gas are available?

   b. What if the tanks are $\frac{1}{8}$ full?

   c. What if the tanks are $\frac{5}{8}$ full?

Name ______________________ Date ______________

Decimals, Fractions, and Division (Page 1 of 1)

1. Ms. Pettigrew coaches the chess club. At the end of the year, eight members of the club bought her a new chess set as a present. If the chess set cost $6, how much did each member have to pay? Explain how you solved this problem.

2. Six musicians decided to play together on the street corner. They earned a total of $4 and decided to share the money equally. Joan and Juan wanted to find out how much each musician would get. They each solved the problem differently.

Would you have solved this problem like Joan did, like Juan did, or in another way? Why?

3. Complete the table on the right. You may use your calculator.

	0.8
15 ÷ 10	

Name ______________________ Date ______________

Checking the Bill (Page 1 of 1)

1. In a restaurant, Patricia received the bill on the right:

   After quickly checking the bill, Patricia discovered that the addition was wrong. Explain how she could have known that the addition was wrong without adding up the whole bill.

	$6.98
	$7.98
	$4.25
	$6.96
	$2.00
	$3.00
Total	$41.17

2. Below are some receipts from shopping trips to the supermarket. Roughly estimate the total for each receipt and explain how you estimated.

a.
$3.98
$3.98
$3.98
$3.98
$3.98

Total ______

b.
$2.97
$3.06
$5.99
$0.99

Total ______

c.
$0.75
$0.75
$0.75
$0.75
$0.75
$0.75
$0.75
$0.75

Total ______

d.
$0.24
$0.24
$0.24
$0.24
$0.24
$0.24
$0.24

Total ______

e.
$5.96
$2.96
$4.96
$3.96
$5.96

Total ______

f.
$5.98
$9.02
$6.97
$3.03
$4.02
$2.98

Total ______

g.
$3.31
$2.58
$6.49
$2.65
$5.38
$0.44

Total ______

h.
$6.38
$4.12
$9.72
$4.08
$5.74
$2.68

Total ______

Name ______________________ Date ______________

Collecting Pennies (Page 1 of 1)

To the right is a picture of a penny-collecting tube. According to the scale on the tube, the tube is full when there are 100 pennies ($1.00) in it.

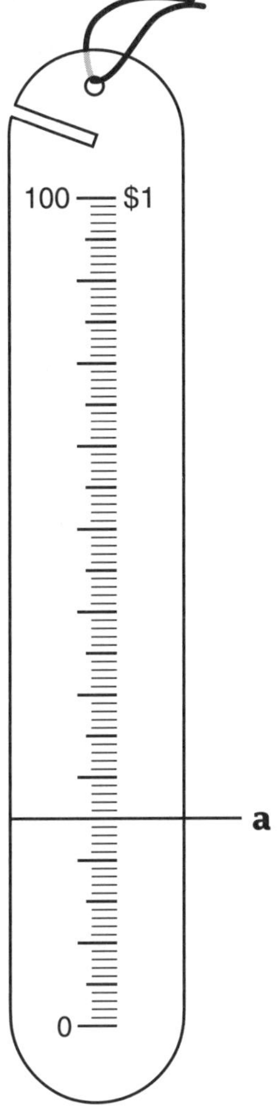

1. Show how full the tube would be for each of the following amounts of money. You can do this by drawing a line on the tube and writing the letter for the problem next to the line. Problem **a** has been done for you.

a. $0.25 **b.** $0.50 **c.** $0.75

d. $0.20 **e.** $0.02 **f.** $0.77

g. $0.40 **h.** $0.80 **i.** $0.98

j. $0.09 **k.** $1.00 **l.** $0.10

m. $0.67

2. Draw a line connecting each decimal given below to its place on the number line.

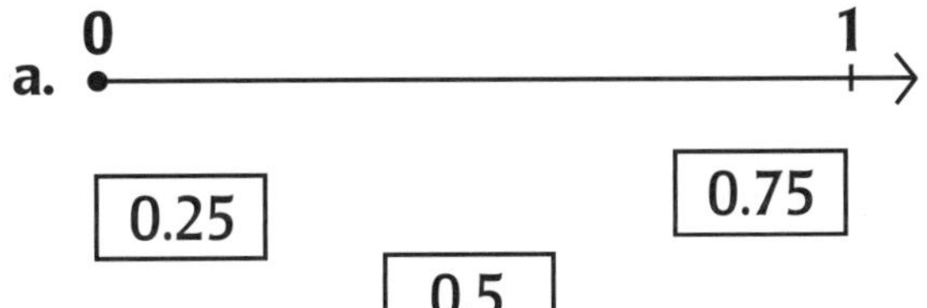

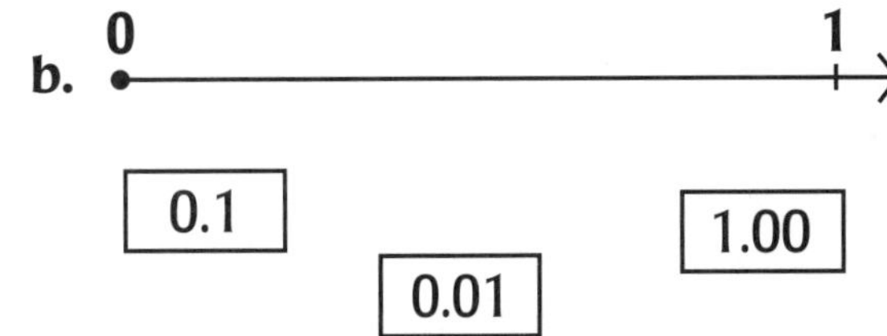

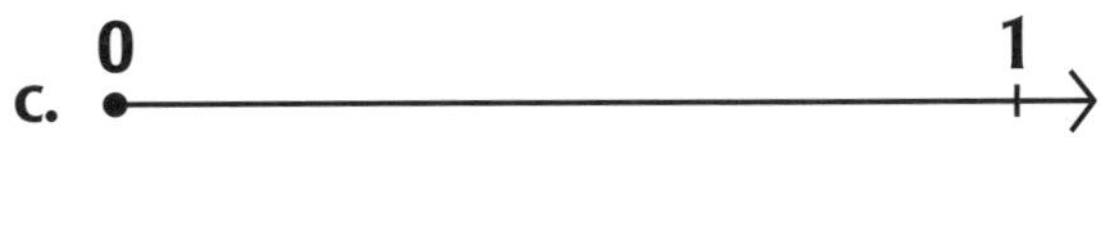

0.24 0.72 0.48

d. 0 1

0.666666 0.33 0.6

Name ______________________ Date ______________________

Section G. Decimals

How Far from School (Page 1 of 1)

School 1 km 2 km 3 km

1. Karen, Bob, Els, Martin, and Tonya all live on the road that passes by their school. Above is a map of the road, with distances from the school indicated in kilometers. According to the following information, mark each person's home on the map and label the mark with the letter a, b, c, d, or e.

   a. Karen lives 1.75 kilometers from school.

   b. Bob lives $2\frac{1}{4}$ kilometers from school.

   c. Els lives 2.05 kilometers from school.

   d. Martin lives 0.33 kilometers from school.

   e. Tonya lives $1\frac{2}{3}$ kilometers from school.

2. Draw a line connecting each decimal or fraction below to its place on the number line.

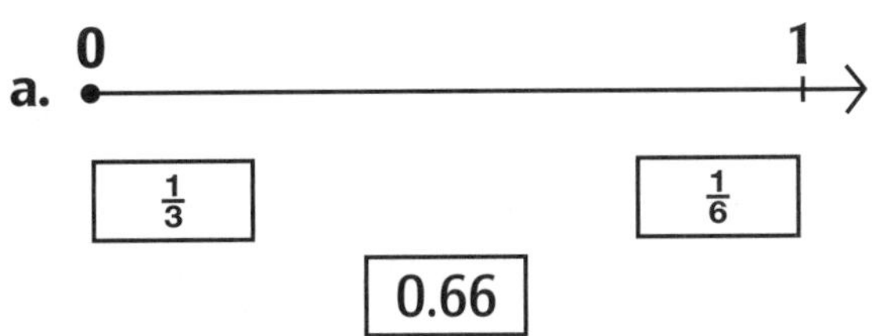

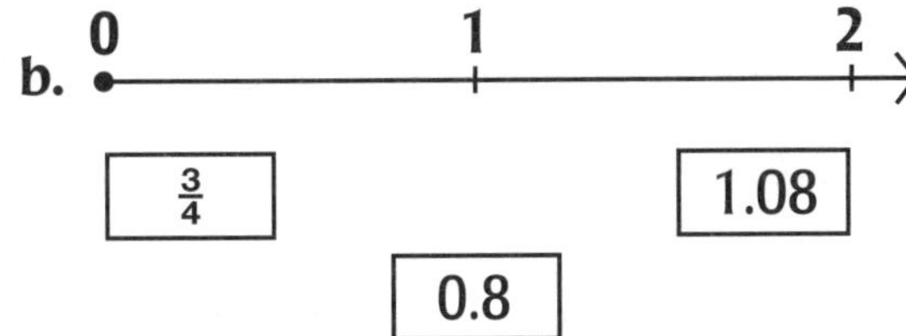

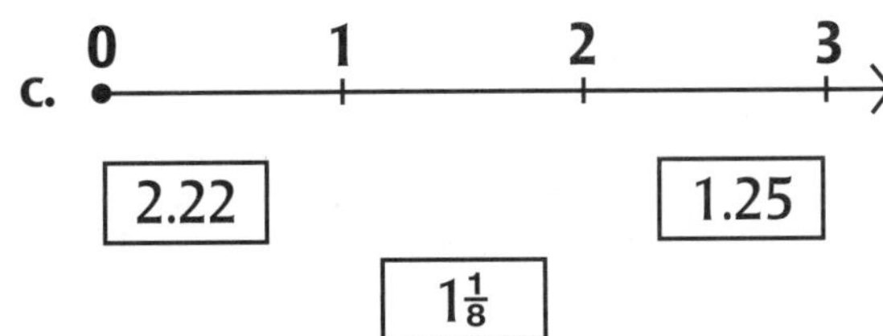

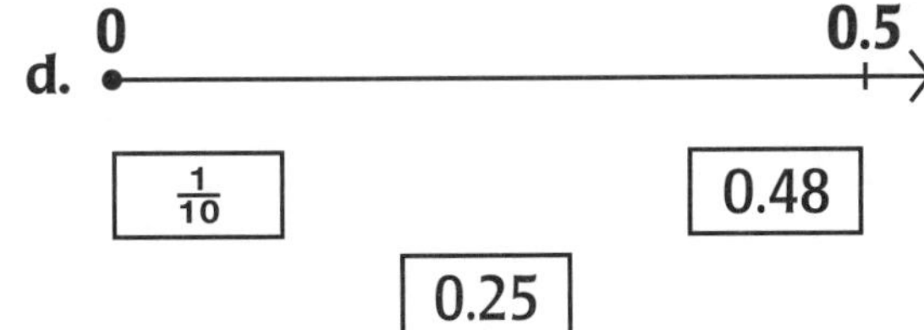

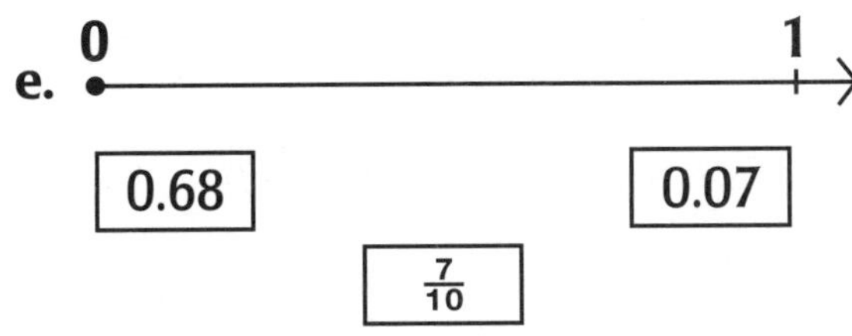

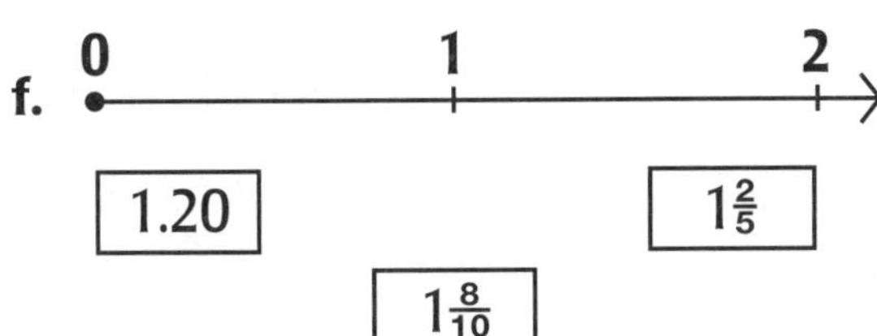

Name ______________________ Date ______________

Section G. Decimals

The Metric System (Page 1 of 2)

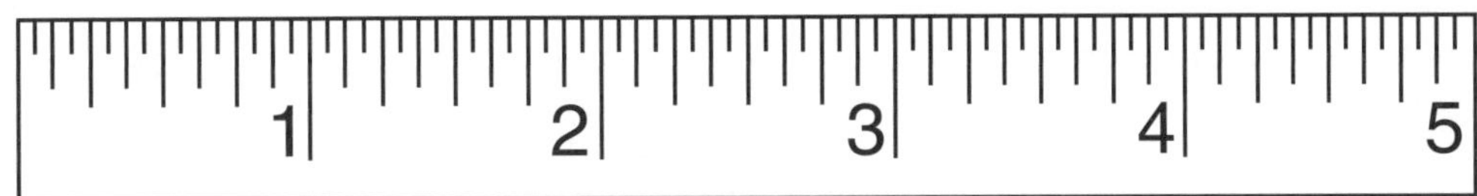

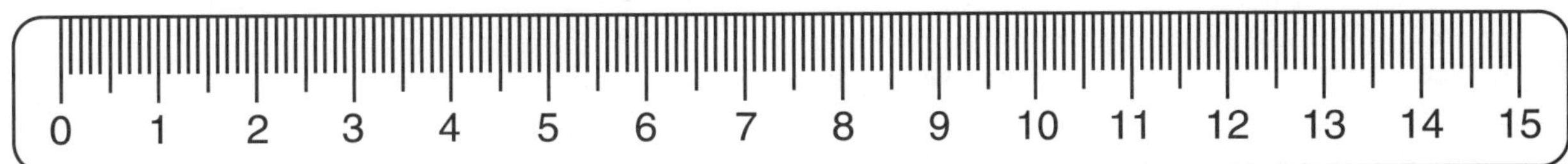

1. Two different rulers are shown above. One of them is marked with inches and the other with centimeters. Which one—the first or the second—shows centimeters? How can you tell?

2. When you look at the two rulers above, you can see that one is metric and the other is not. What characteristic does the metric system have that helps you to determine which ruler is metric?

3. How long is one meter? Look around you. Is there anything in the classroom that is about one meter long or wide or high?

Millimeters, centimeters, decimeters, and ***meters*** are metric units used to measure length.

Milliliters, centiliters, deciliters, and ***liters*** are metric units used to measure volume.

Name ______________________________ Date ______________

The Metric System (Page 2 of 2)

The prefixes before the words *meter* and *liter* come from Latin words. Some of them are defined below:

milli: one-thousandth	one millimeter = one-thousandth of a meter
centi: one-hundredth	one centimeter = one-hundredth of a meter
deci: one-tenth	one decimeter = one-tenth of a meter

There are also prefixes for larger measurement units:

kilo: one thousand	one kilometer = one thousand meters
hecto: one hundred	one hectometer = one hundred meters
deca: ten	one decameter = ten meters

A millimeter is one-thousandth of a meter. There are two ways to write this:

as a fraction:	1 millimeter = $\frac{1}{1000}$ meter
as a decimal:	1 millimeter = 0.001 meter

4. Describe each of the following relationships in two ways: with a fraction and with a decimal.

a. 1 centimeter = ______ meter

1 centimeter = ______ meter

b. 1 decimeter = ______ meter

1 decimeter = ______ meter

c. 1 meter = ______ kilometer

1 meter = ______ kilometer

d. 1 meter = ______ hectometer

1 meter = ______ hectometer

e. 1 meter = ______ decameter

1 meter = ______ decameter

Name ______________________ Date ____________

A Little about Liters (Page 1 of 1)

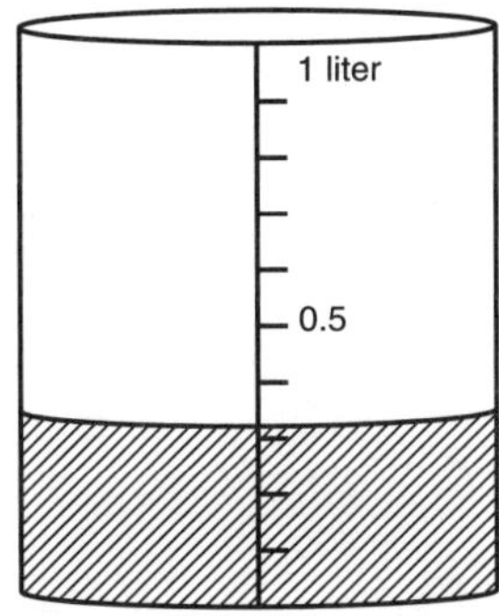

Graduated cylinders like the one on the left are used to measure volume. A full cylinder contains one liter.

1. What volume of liquid is in the cylinder on the left?

2. What interval markings would you want on the cylinder if you had to measure 0.07 liter of liquid?

Milliliters, centiliters, deciliters, and ***liters*** are metric units used to measure liquid volume. They are defined below:

milli: one-thousandth	one milliliter = $\frac{1}{1000}$ liter
centi: one-hundredth	one centiliter = $\frac{1}{100}$ liter
deci: one-tenth	one deciliter = $\frac{1}{10}$ liter

3. Mark the level on the graduated cylinder on the right to show 0.09 liter of liquid.

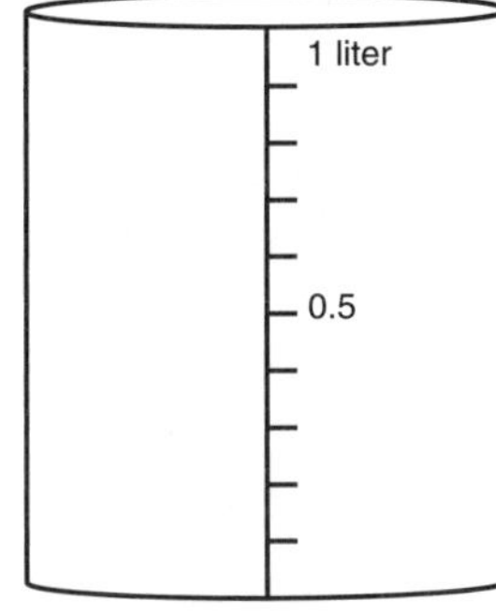

4. a. How much is 0.09 liter in deciliters?
0.09 liter = _____ deciliters

b. How much is 0.09 liter in centiliters?
0.09 liter = _____ centiliters

c. How much is 0.09 liter in milliliters?
0.09 liter = _____ milliliters

5. a. Draw the level of the liquid in each cylinder to show the given amount.

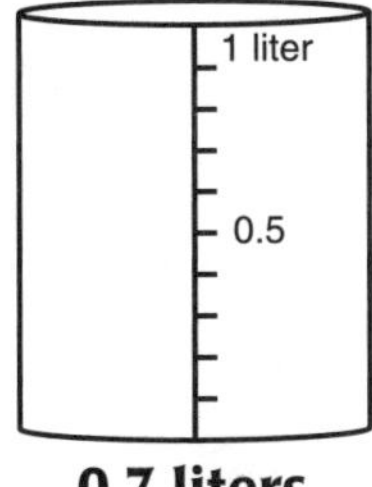

0.7 liters

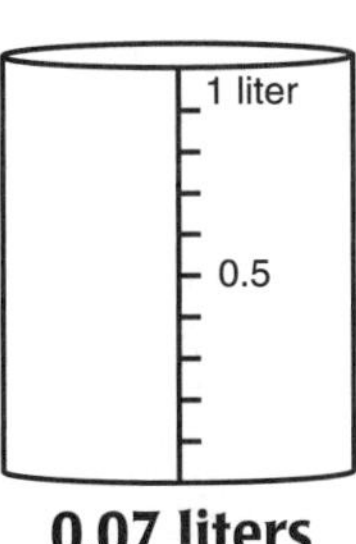

0.07 liters

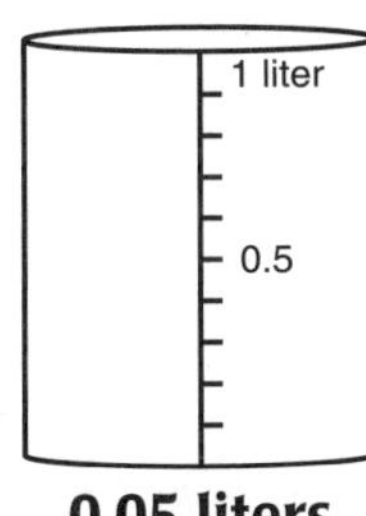

0.05 liters

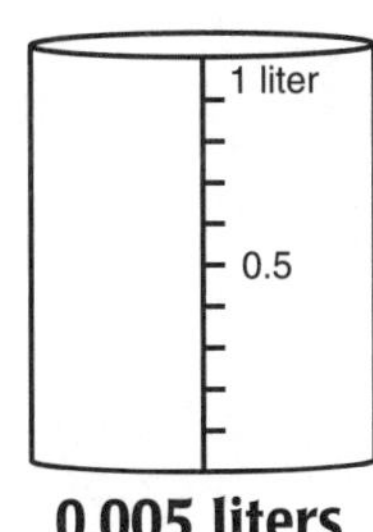

0.005 liters

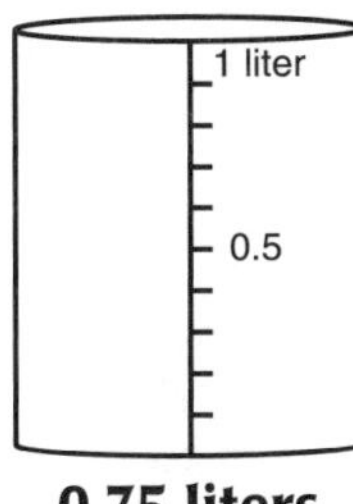

0.75 liters

b. Order the amounts of liquid shown in part **a** from smallest to largest.

Name ______________________________ Date ______________

How Much Did You Pour? (Page 1 of 1)

Below you see 1-liter graduated cylinders. Draw the levels of liquid in the first and second cylinders. Then draw the level of liquid in the third cylinder that will be a combination of the levels in the first two cylinders. Below the third cylinder, write the amount of liquid that it contains.

1.

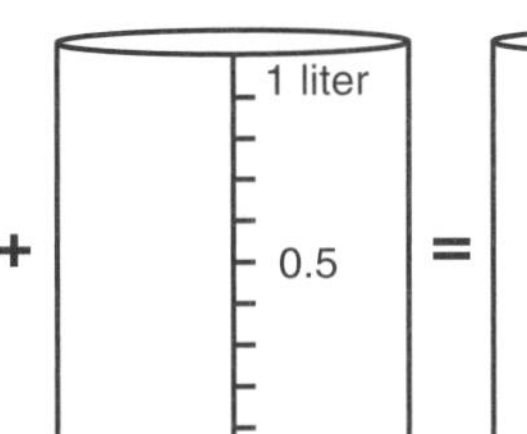

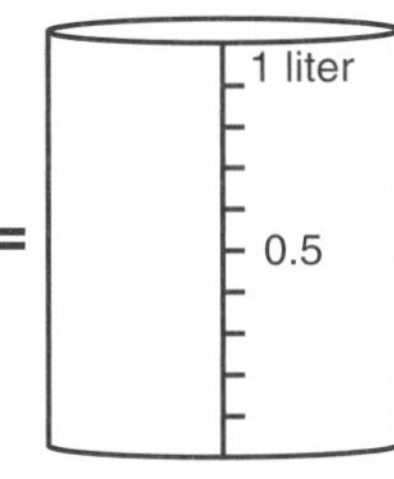

0.3 liters + 0.5 liters = ____ liters

2.

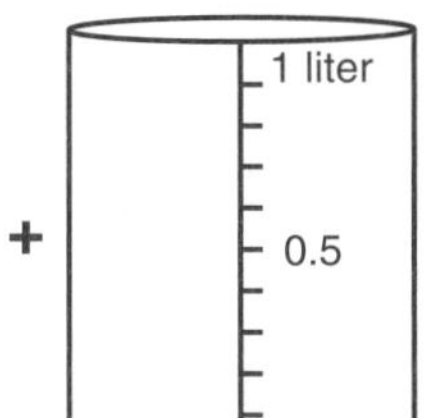

0.9 liters + 0.09 liters = ____ liters

3.

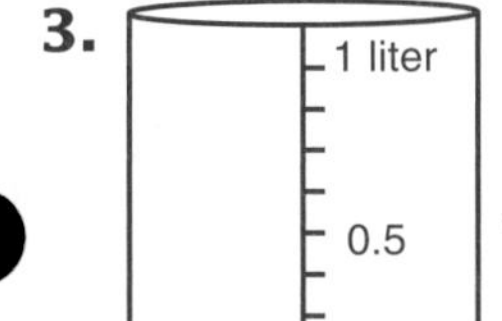

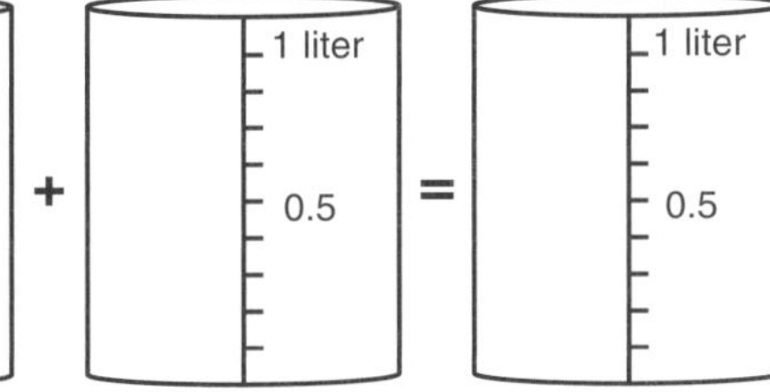

0.25 liters + 0.6 liters = ____ liters

4.

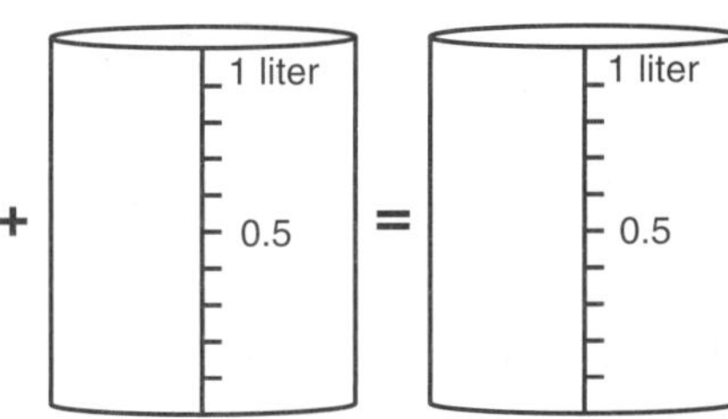

0.375 liters + 0.5 liters = ____ liters

5.

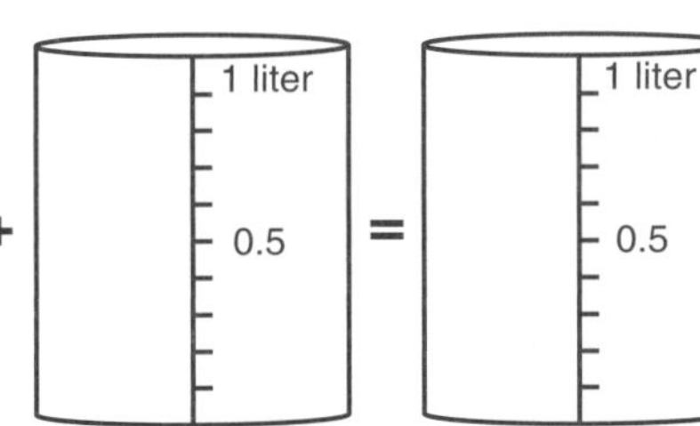

0.4 liters + 0.075 liters = ____ liters

6.

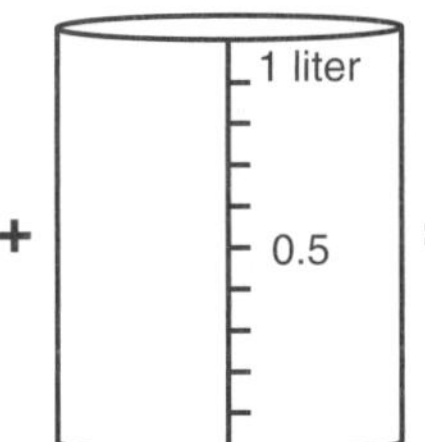

0.1 liters + 0.805 liters = ____ liters

Section H.
Percents

Name ______________________ Date ______________

Class Attendance (Page 1 of 1)

French Class

1. Above you see two classes, French and Spanish. David claims that the French class is better attended than the Spanish class. What does he mean? Could he be correct?

2. Fill in the ratio table below to show larger and smaller classes that are just as well attended as the French class.

Class	French	Chinese	Math	Art	Chorus	Band	Science
Students	8		10				
Total Desks	12	60		3	45		

3. a. What fraction represents the attendance of the French class?

   b. What percent represents the attendance of the French class?

4. To the right you see a history class. Use the ratio table below to show other classes that have the same attendance as the history class.

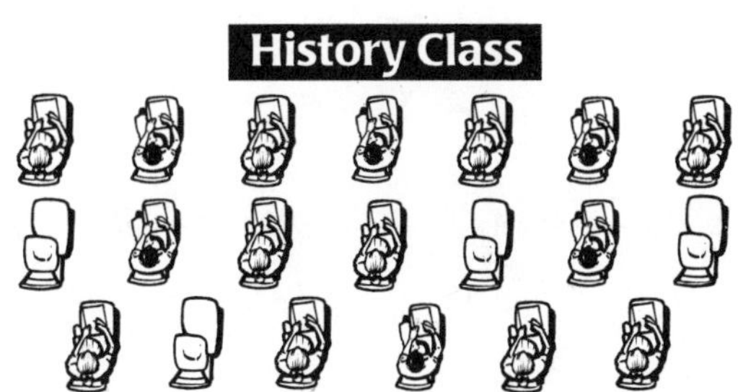

Class	History	Latin	German	Sewing	Gym	Biology
Students						
Total Desks						

5. Which class is better attended, French or history? Explain your answer.

Name ______________________ Date ______________

Going to the Movies (Page 1 of 1)

Theater 1

Theater 2

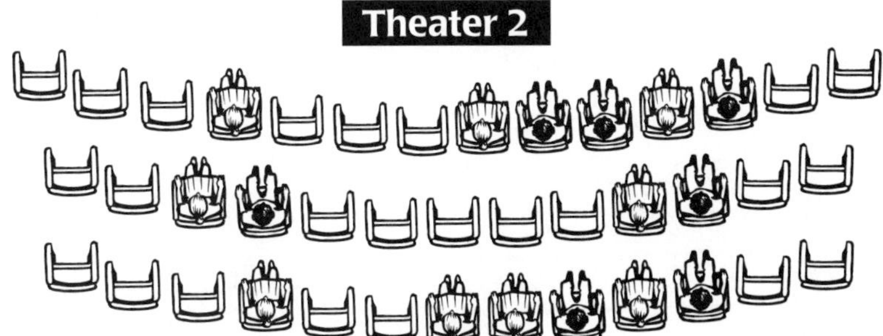

1. Pictured above are two movie theaters. The theater on the left is showing *Revenge of the Canadian Geese*. The theater on the right is showing *Betty, the Monster Cockroach*. Which of the two movies is more popular? Explain.

2. Theater 1 does not have many people in it, but it is also a very small theater. Fill in the ratio table below to show larger and smaller theaters with the same attendance ratio as Theater 1 (T1).

Theater	T1	T3	T4	T5	T6	T7	T8	T9	T10
Seats Occupied	6		1						
Total Seats	24	12		20					100

3. Theater 2 has more people in it than Theater 1, and it is a larger theater. Use the ratio table below to show larger and smaller theaters with the same attendance ratio as Theater 2 (T2).

Theater	T2	T11	T12	T13	T14	T15	T16	T17	T18
Seats Occupied	16								
Total Seats	40								100

4. **a.** What fraction of T1 is occupied? T2? **b.** What percent of T1 is occupied? T2?

   c. Shade the two bars below to show the attendance at T1 and T2.

T1 0 — 24; 0 % — 100 %

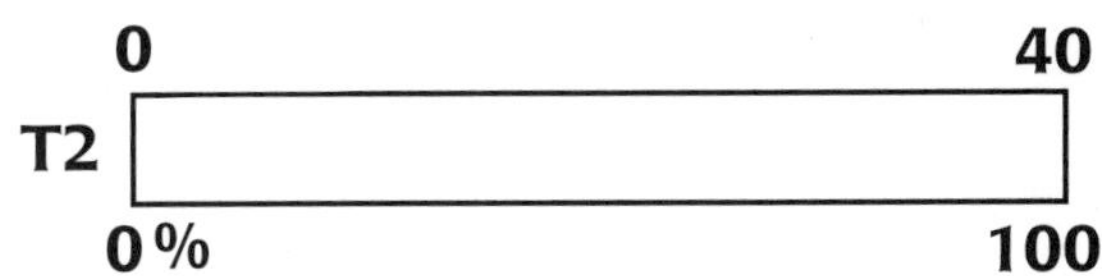

5. Which movie would you rather see—*Betty, the Monster Cockroach* or *Revenge of the Canadian Geese*? Why?

Name ______________________________ Date ______________

School Plays (Page 1 of 1)

The High Drama Club produced seven plays last year in different places around town. Some of the plays were more popular than others. The president of the club kept track of attendance using the percent bars shown below. For each play, express the attendance as a fraction, as a percent, and as the number of seats occupied.

1.

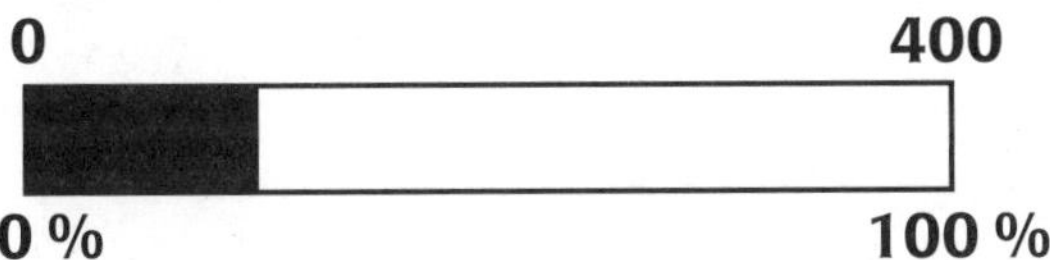

Fraction occupied: __________

Percent occupied: __________

Seats occupied: __________

2.

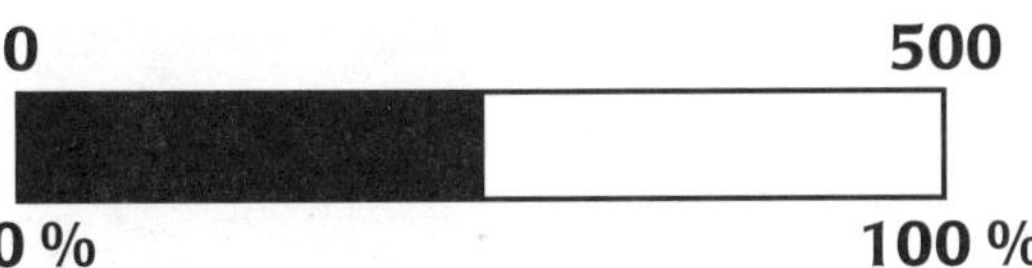

Fraction occupied: __________

Percent occupied: __________

Seats occupied: __________

3.

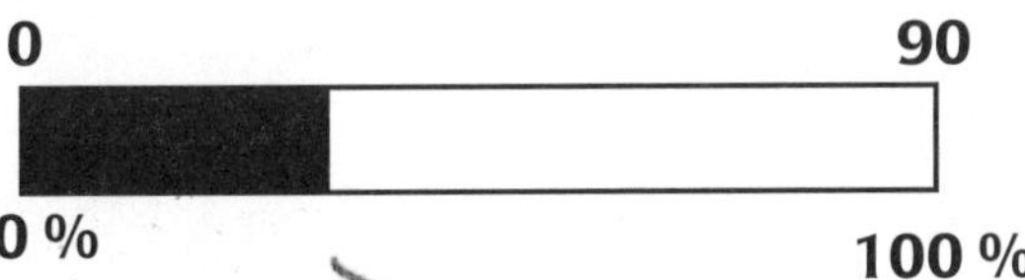

Fraction occupied: __________

Percent occupied: __________

Seats occupied: __________

4.

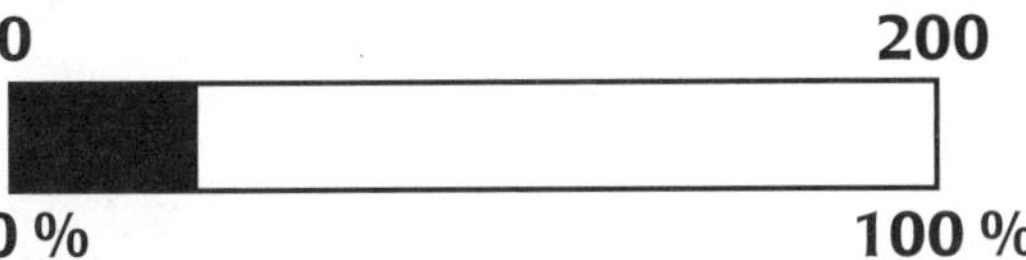

Fraction occupied: __________

Percent occupied: __________

Seats occupied: __________

5.

Fraction occupied: __________

Percent occupied: __________

Seats occupied: __________

6.

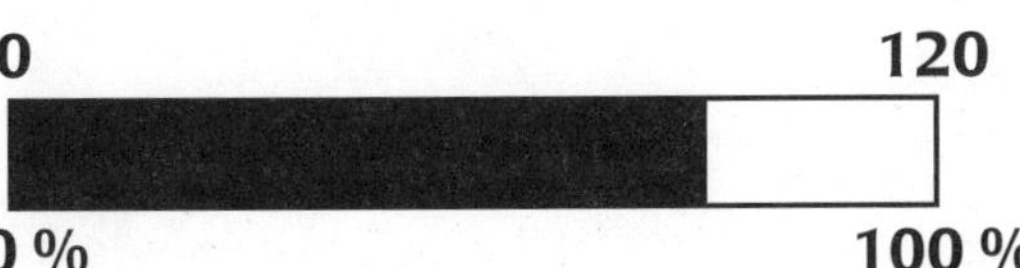

Fraction occupied: __________

Percent occupied: __________

Seats occupied: __________

7.

Fraction occupied: __________

Percent occupied: __________

Seats occupied: __________

Name ______________________ Date ______________

East vs. West (Page 1 of 2)

There is a great rivalry between the sports teams of East Middle School and West Middle School.

1. At the last gymnastics meet between the two schools, the ratio of East fans to West fans was 13 to 7. The sports writer for the East Middle School newspaper wanted to report the percentages of fans for each school.

   a. Fill in the ratio table below to find out how many East and West fans there would be in 100 spectators.

East Fans	13				
West Fans	7				
Total	20				

   b. Shade and label the following percent bars to show the percentages of East fans and West fans.

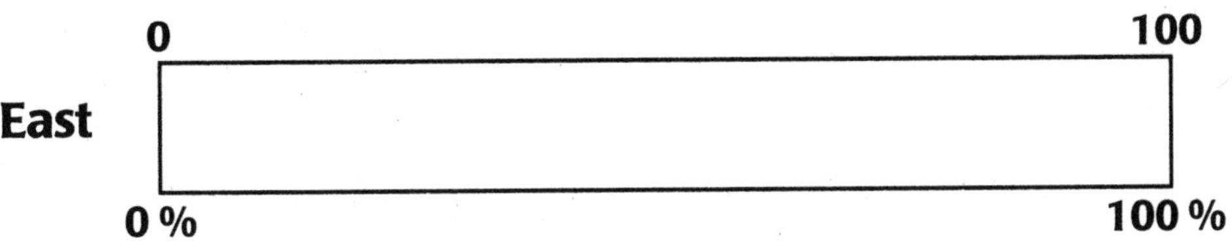

   c. There were 180 fans at the gymnastics meet. Use either a percent bar or a ratio table to find the number of fans for each team.

East vs. West (Page 2 of 2)

2. At the last swim meet between East and West, the ratio of fans was 4 East fans to 6 West fans.

a. What percent represents the number of fans for East? for West?

b. There were 65 fans at the meet. How many fans did each team have? Show your strategy.

3. At one basketball game between East and West, there were 19 East fans for every 13 West fans.

a. Approximately what percents represent the numbers of East and West fans?

b. There were 2,500 fans at the game. Approximately how many fans did each team have? Show your strategy.

4. At which of the sporting event mentioned in problems **1–3** did East Middle School have the highest percentage of fans?

5. At which sporting event did East Middle School have the most fans?

Name ______________________ Date ______________

Tipping (Page 1 of 1)

1. a. Ms. Eng usually gives a 10% tip in restaurants. Fill in the 10% column in the table below to show what Ms. Eng would leave for a tip.

b. Mr. Lonetree usually gives a 15% tip in restaurants. Fill in the 15% column in the table below to show what Mr. Lonetree would leave for a tip.

Restaurant	Bill	10% Tip	15% Tip
Hamburger Heaven	\$13.65		
The Apple Dumplin' Diner	\$ 7.42		
Frank's Pizza Palace	\$29.10		
The Newton Grill	\$52.85		
Chez Louis	\$77.50		

2. Decide whether Ms. Eng or Mr. Lonetree left the given tip for each of the following bills. Explain your choice.

Total Bill **Tip**

a. \$ 32.08   

b. \$ 27.27   

c. \$ 8.32   

d. \$14.00  

e. \$19.58   

Section I.
Arrow Language

Name ______________________ Date ______________

Two Steps (Page 1 of 1)

1. Find the answers to the following problems without using a pencil and paper or a calculator:

   a. 547 + 99 =
   b. 437 + 99 =
   c. 8,035 + 99 =
   d. 63 + 99 =
   e. 21,653 + 99 =

2. How did you find the answers to problem **1?** Did you discover any shortcuts?

Karen has $483 in her savings account. She deposits another $90 into her account and then figures out her total balance.

The following arrow string shows Karen's method:

$$483 \xrightarrow{+100} 583 \xrightarrow{-10} 573$$

3. Rewrite the following problems as arrow strings and then solve them. Each arrow string should show how to use Karen's method to mentally calculate the answer.

   a. 624 + 99
   b. 624 − 99
   c. 5,444 + 999
   d. 5,444 − 999
   e. 832 + 90
   f. 832 − 90
   g. 1,573 + 98
   h. 1,573 − 98
   i. 365 + 997
   j. 4,526 − 997
   k. 6,000 − 991
   l. 5,001 + 998

Name ______________________________ Date ______________

Different Ways (Page 1 of 1)

Abraham and Beth want to add 235 and 48 without using a calculator or a pencil and paper. Here is how each one solves the problem mentally.

1. Write one arrow string showing Abraham's method and another one showing Beth's method.

2. Using arrow strings, describe at least three ways to mentally calculate 492 + 39.

3. Use arrow strings to describe at least two ways to mentally calculate each of the following problems. Be sure to include your answers.

a. 468 + 29

b. 986 − 91

c. 99 + 250

d. 986 − 49

e. 328 + 28

f. 506 + 58

g. 880 − 28

h. 640 − 48

i. 543 + 39

j. 3,962 + 39

Name ________________________________ Date ________________

Winning and Losing (Page 1 of 1)

Every day after school, Jesse plays marbles. Yesterday he started the day with 136 marbles and won 16 more. Today he lost 9 marbles.

Each arrow string below shows a method for calculating the number of marbles that Jesse has now.

$$136 \xrightarrow{+16} 152 \xrightarrow{-9} 143$$

$$136 \xrightarrow{+7} 143$$

1. Explain why both arrow strings above show the correct number of marbles that Jesse has now.

2. Fill in the missing numbers for each of the following arrow strings and then shorten each string so that it has only one arrow:

a. $938 \xrightarrow{+25}$ _____ $\xrightarrow{-20}$ _____

b. $589 \xrightarrow{-100}$ _____ $\xrightarrow{+99}$ _____

c. $97 \xrightarrow{+1{,}000}$ _____ $\xrightarrow{+1{,}000}$ _____

d. $35 \xrightarrow{+1{,}000}$ _____ $\xrightarrow{-800}$ _____

e. $763 \xrightarrow{-98}$ _____ $\xrightarrow{-2}$ _____

f. $603 \xrightarrow{+75}$ _____ $\xrightarrow{+25}$ _____

g. $800 \xrightarrow{+98}$ _____ $\xrightarrow{-100}$ _____

h. $800 \xrightarrow{-100}$ _____ $\xrightarrow{+98}$ _____

3. The arrow strings for problems **2g** and **2h** can be rewritten as the same string. Explain why.

4. The following arrow strings do not produce the same outcome even though they contain the same numbers and operations. Explain why.

$600 \xrightarrow{\times 2}$ _____ $\xrightarrow{+200}$ _____

$600 \xrightarrow{+200}$ _____ $\xrightarrow{\times 2}$ _____

Name ______________________ Date ____________

Multiplication and Division (Page 1 of 1)

Mr. Starks has an aquarium in his classroom. In order to find its volume, Mr. Starks's students first determine the aquarium's dimensions, as shown on the left. Maya, Luisa, and Thomas each propose a different string to find the aquarium's volume:

$60 \xrightarrow{\times 40} 2{,}400 \xrightarrow{\times 50} 120{,}000 \text{ cm}^3$

$60 \xrightarrow{\times 50} 3{,}000 \xrightarrow{\times 40} 120{,}000 \text{ cm}^3$

$60 \xrightarrow{\times 2{,}000} 120{,}000 \text{ cm}^3$

1. Compare the above three arrow strings. Do all three strings correctly show how to find the aquarium's volume? Explain.

2. For each of the following arrow strings, fill in the missing numbers. Then write another arrow string that shows an alternative way to calculate the answer.

a. $8 \xrightarrow{\times 5}$ _____ $\xrightarrow{\times 4}$ _____

b. $32 \xrightarrow{\times 2}$ _____ $\xrightarrow{\times 5}$ _____

c. $50 \xrightarrow{\times 8}$ _____ $\xrightarrow{\div 4}$ _____

d. $750 \xrightarrow{\div 3}$ _____ $\xrightarrow{\times 2}$ _____

e. $1{,}050 \xrightarrow{\div 5}$ _____ $\xrightarrow{\div 2}$ _____

f. $9 \xrightarrow{\times 30}$ _____ $\xrightarrow{\div 30}$ _____

g. $12 \xrightarrow{\times 100}$ _____ $\xrightarrow{\times 5}$ _____

3. Compare the following two arrow strings. Why are their outcomes different?

$60 \xrightarrow{\times 5}$ _____ $\xrightarrow{+ 40}$ _____

$60 \xrightarrow{+ 40}$ _____ $\xrightarrow{\times 5}$ _____

Name ______________________ Date ______________

Going Backwards (Page 1 of 1)

1. Find the result of each of the following arrow strings:

a. $38 \xrightarrow{+2} \_\_\_\_ \xrightarrow{\times 4} \_\_\_\_ \xrightarrow{-20} \_\_\_\_ \xrightarrow{\div 2} \_\_\_\_$

b. $70 \xrightarrow{+50} \_\_\_\_ \xrightarrow{-60} \_\_\_\_ \xrightarrow{\times 3} \_\_\_\_ \xrightarrow{-10} \_\_\_\_$

c. $5 \xrightarrow{\times 20} \_\_\_\_ \xrightarrow{-20} \_\_\_\_ \xrightarrow{\times 2} \_\_\_\_ \xrightarrow{\div 2} \_\_\_\_$

d. $606 \xrightarrow{+14} \_\_\_\_ \xrightarrow{\times 2} \_\_\_\_ \xrightarrow{-100} \_\_\_\_ \xrightarrow{+50} \_\_\_\_$

e. $1{,}000 \xrightarrow{\div 4} \_\_\_\_ \xrightarrow{\times 4} \_\_\_\_ \xrightarrow{-500} \_\_\_\_ \xrightarrow{+500} \_\_\_\_$

2. In each of the following arrow strings, the result is given. Fill in all of the missing numbers and find the first number for each string.

a. $\_\_\_\_ \xrightarrow{\times 2} \_\_\_\_ \xrightarrow{\div 4} \_\_\_\_ \xrightarrow{-20} \_\_\_\_ \xrightarrow{\times 7} 35$

b. $\_\_\_\_ \xrightarrow{+19} \_\_\_\_ \xrightarrow{\times 2} \_\_\_\_ \xrightarrow{-100} \_\_\_\_ \xrightarrow{-95} 5$

c. $\_\_\_\_ \xrightarrow{+2} \_\_\_\_ \xrightarrow{\times 2} \_\_\_\_ \xrightarrow{-20} \_\_\_\_ \xrightarrow{\div 2} 40$

d. $\_\_\_\_ \xrightarrow{+50} \_\_\_\_ \xrightarrow{-10} \_\_\_\_ \xrightarrow{\div 3} \_\_\_\_ \xrightarrow{-2} 78$

e. $\_\_\_\_ \xrightarrow{+50} \_\_\_\_ \xrightarrow{\div 2} \_\_\_\_ \xrightarrow{-396} \_\_\_\_ \xrightarrow{\times 4} 16$

3. Make up your own arrow strings with the following results and number of arrows:

a. $\_\_\_\_ \longrightarrow \_\_\_\_ \longrightarrow \_\_\_\_ \longrightarrow \_\_\_\_ \longrightarrow 16$

b. $\_\_\_\_ \longrightarrow \_\_\_\_ \longrightarrow \_\_\_\_ \longrightarrow \_\_\_\_ \longrightarrow 20$

c. $\_\_\_\_ \longrightarrow \_\_\_\_ \longrightarrow \_\_\_\_ \longrightarrow \_\_\_\_ \longrightarrow 52$

Name ______________________ Date ______________

The Deli (Page 1 of 1)

Amy is buying some meat and cheese at the deli. Swiss cheese costs $3.60 per pound, but Amy needs only 0.75 pound. She uses the following arrow string to compute how much 0.75 pound of Swiss cheese will cost:

$\$3.60 \xrightarrow{\div 4} \$0.90 \xrightarrow{\times 3} \$2.70$

1. Explain Amy's method.

2. Write an arrow string that describes how to calculate each of the following:

a. Ham costs $3.20 per pound. How much does 0.75 pound cost?

b. Cheddar cheese costs $7.50 per pound. How much does 0.80 pound cost?

c. Salami costs $4.00 per pound. How much does 0.40 pound cost?

d. Bologna costs $3.00 per pound. How much does 0.60 pound cost?

e. Pepperoni costs $2.40 per pound. How much does 1.25 pounds cost?

f. Carol pays $1.80 for 0.75 pound of American cheese. What is the price per pound of American cheese?

g. Andrew pays $1.60 for 0.80 pound of sliced turkey. What is the price per pound of sliced turkey?

h. Jesse pays $3.50 for 1.25 pounds of corned beef. What is the price per pound of corned beef?

Section J.
Focus on Fractions

Name ______________________ Date ______________

Sisters (Page 1 of 2)

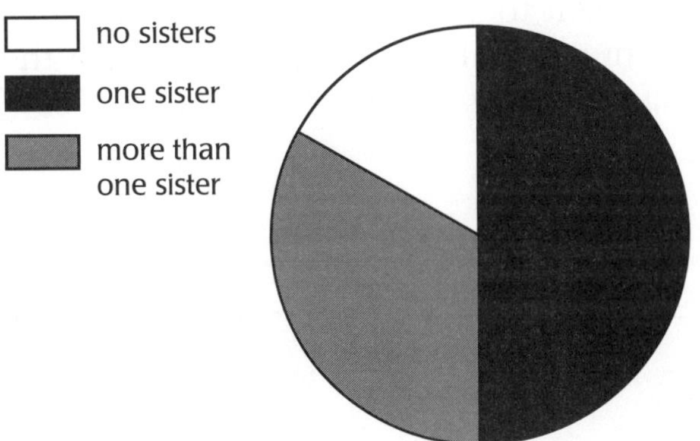

1. Ms. Nakamura's students take a survey and discover that $\frac{1}{2}$ of the students in the class have only one sister and $\frac{1}{3}$ have more than one sister. The class displays the survey results in the pie chart shown on the right.

a. How many students could there be in Ms. Nakamura's class? Explain why there are several possible answers to this question.

b. What fraction of the class has one or more sisters?

c. What fraction of the class has no sisters?

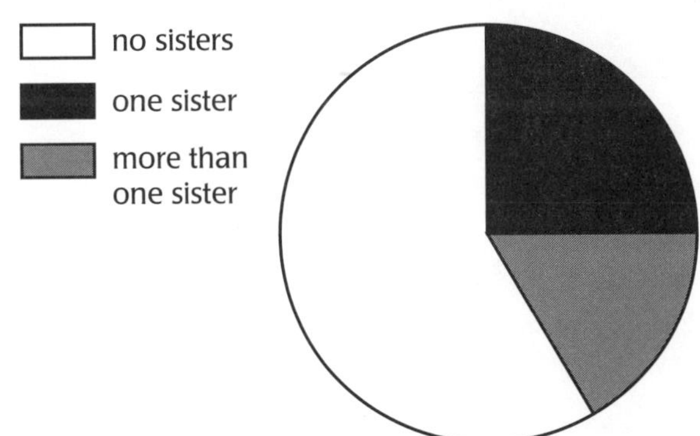

2. Mr. Brown's class conducted the same survey. One-fourth of his students have only one sister, and $\frac{1}{6}$ of the students have more than one sister. The pie chart for Mr. Brown's class is shown on the right.

a. How many students could there be in Mr. Brown's class? Find several possible answers.

b. What fraction of the class has one or more sisters?

c. What fraction of the class has no sisters?

Name ____________________ Date ____________

Sisters (Page 2 of 2)

If you want to add fractions with different denominators, you can make bars, each with the same number of segments, to represent the fractions. For example, $\frac{1}{4} + \frac{1}{6}$ can be solved by using two bars with 12 segments each, as shown below.

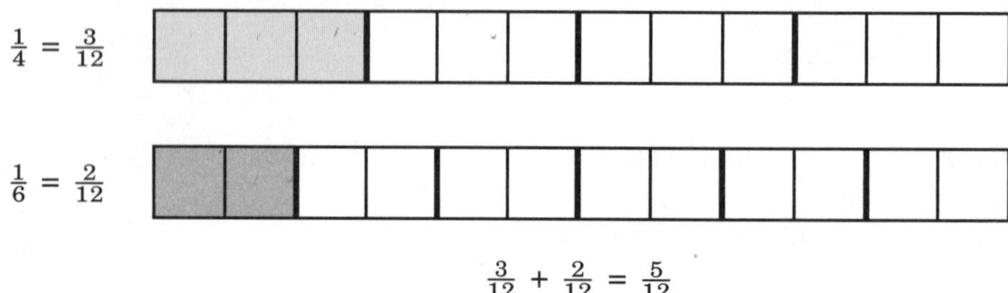

$$\frac{3}{12} + \frac{2}{12} = \frac{5}{12}$$

3. Explain why you could also use two bars with 24 segments each to add $\frac{1}{4}$ and $\frac{1}{6}$.

4. Solve the following addition problems. If necessary, make two bars, each with the same number of segments, to represent the two fractions.

a. $\frac{1}{4} + \frac{1}{8} =$

b. $\frac{2}{3} + \frac{1}{4} =$

c. $\frac{1}{2} + \frac{1}{6} =$

d. $\frac{1}{6} + \frac{4}{9} =$

e. $\frac{1}{10} + \frac{3}{4} =$

f. $\frac{3}{8} + \frac{1}{2} =$

g. $\frac{1}{3} + \frac{1}{5} =$

h. $\frac{2}{6} + \frac{2}{4} =$

i. $\frac{2}{4} + \frac{3}{10} =$

Name ______________________ Date ______________

Comparing Two Schools (Page 1 of 2)

Below is some information about the students at Greenfield and Brendel middle schools.

I. Female Students

Greenfield: $\frac{2}{3}$ of the students

Brendel: $\frac{5}{8}$ of the students

II. Students Transported by Bus

Greenfield: $\frac{5}{6}$ of the students

Brendel: $\frac{3}{4}$ of the students

III. Sixth-Grade Students

Greenfield: $\frac{1}{4}$ of the students

Brendel: $\frac{2}{5}$ of the students

IV. Sports Club Members

Greenfield: $\frac{1}{4}$ of the students

Brendel: $\frac{3}{10}$ of the students

1. a. Shade each bar above so that it represents the fraction of students given.

b. For each category listed above, divide the pair of bars into the same number of segments so that you can compare the fractions.

c. For each category, determine which school has the larger fraction of students.

Name ______________________ Date ____________

Comparing Two Schools (Page 2 of 2)

2. Compare the following pairs of fractions and circle the larger fraction. If necessary, make two bars to represent the two fractions.

   a. $\frac{1}{4}$ and $\frac{1}{5}$ **b.** $\frac{2}{3}$ and $\frac{6}{8}$ **c.** $\frac{2}{6}$ and $\frac{4}{9}$

   d. $\frac{1}{3}$ and $\frac{2}{5}$ **e.** $\frac{2}{3}$ and $\frac{1}{2}$ **f.** $\frac{3}{8}$ and $\frac{1}{4}$

   g. $\frac{3}{4}$ and $\frac{4}{5}$ **h.** $\frac{3}{5}$ and $\frac{3}{4}$ **i.** $\frac{4}{9}$ and $\frac{1}{3}$

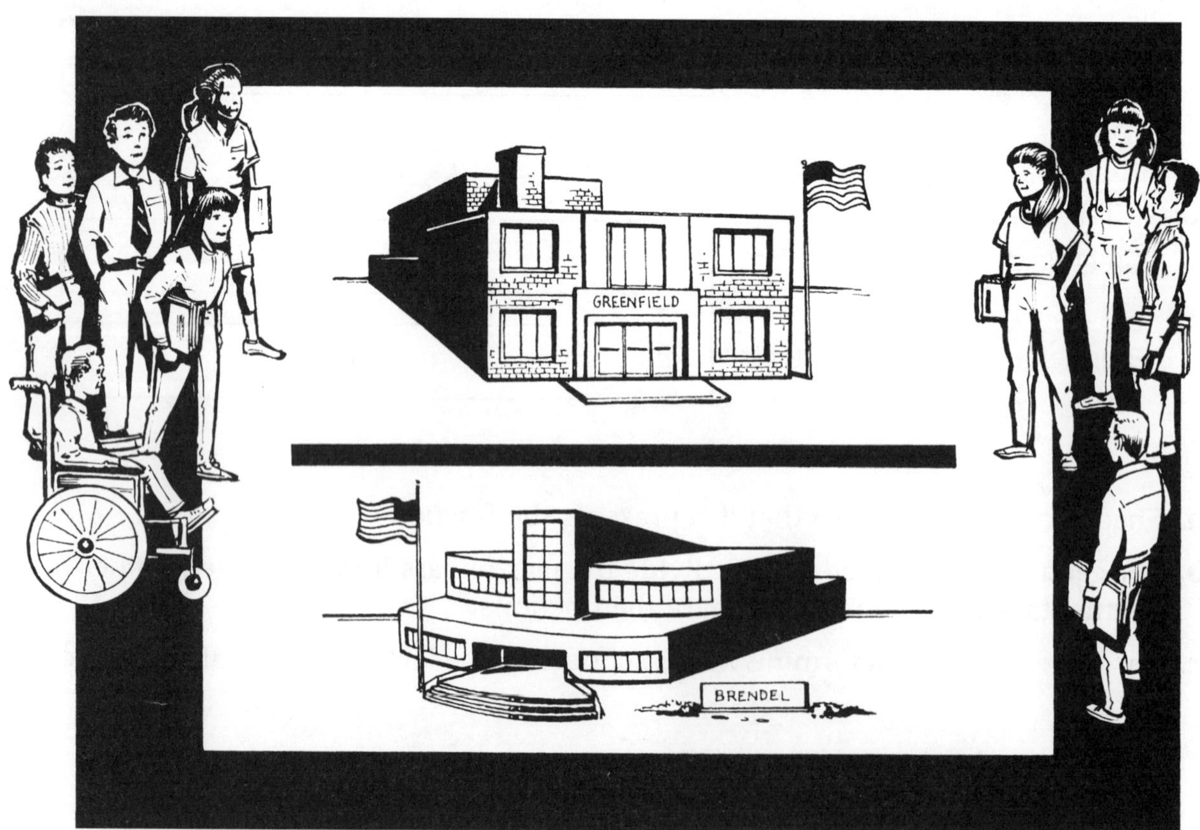

Name ______________________ Date ______________

What Difference? (Page 1 of 1)

Among Ms. Washington's students, $\frac{2}{3}$ of the class participate in a sport. One-fourth of the class play basketball. Ms. Washington asks her students to calculate what fraction of the class participates in a sport other than basketball. Thomas and Jane each use a different method to solve the problem.

IF THE CLASS HAS 24 STUDENTS, THEN 16 OF THEM PLAY SPORTS AND 6 OF THEM PLAY BASKETBALL. THIS MEANS THAT 10 PLAY ANOTHER SPORT. 10 OUT OF 24 IS THE SAME AS 5 OUT OF 12, SO THE ANSWER IS $\frac{5}{12}$.

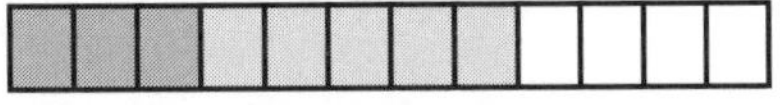

I HAVE DRAWN A BAR WITH 12 SEGMENTS. $\frac{2}{3}$ OF 12 SEGMENTS IS 8 SEGMENTS. $\frac{1}{4}$ OF 12 SEGMENTS IS 3 SEGMENTS. THE DIFFERENCE IS 5 SEGMENTS, SO THE ANSWER IS $\frac{5}{12}$.

1. In Mr. Guilford's class, $\frac{3}{4}$ of the students participate in a sport. If $\frac{1}{3}$ of the students play basketball, what fraction of the class participates in a sport other than basketball?

2. At Jefferson Middle School, $\frac{1}{3}$ of the students study a foreign language. If $\frac{2}{9}$ of the students study Japanese, what fraction of the students studies a foreign language other than Japanese?

3. Solve the following subtraction problems:

a. $\frac{3}{8} - \frac{1}{4} =$ **b.** $\frac{5}{8} - \frac{2}{4} =$ **c.** $\frac{1}{4} - \frac{1}{6} =$

d. $\frac{4}{5} - \frac{2}{3} =$ **e.** $\frac{2}{3} - \frac{2}{5} =$ **f.** $\frac{6}{8} - \frac{2}{3} =$

g. $\frac{2}{3} - \frac{1}{2} =$ **h.** $\frac{4}{8} - \frac{3}{9} =$ **i.** $\frac{4}{9} - \frac{2}{6} =$

Name ______________________ Date __________

Section J. Focus on Fractions

Cleaning Up (Page 1 of 1)

1. Groups of students from Parker School have been helping to clean up city parks. The students are able to earn money by recycling the glass bottles and aluminum cans that they collect. If the students in each group share the money equally, how much money will each student earn in each of the following situations? Write your answers in two ways: as fractions and as dollar amounts. Do not use your calculator.

   a. \$4 earned by five students

   b. \$5 earned by four students

   c. \$3 earned by four students

   d. \$3 earned by eight students

   e. \$6 earned by four students

   f. \$6 earned by eight students

2. If five people earn one dollar, each person gets $\frac{1}{5}$ of a dollar, or 20 cents. This amount can also be written as \$0.20. How does your calculator display $\frac{1}{5}$ of a dollar?

3. Use your calculator to compute the following amounts of money:

   a. $\frac{1}{8}$ of a dollar

   b. $\frac{1}{3}$ of a dollar

   c. $\frac{1}{4}$ of a dollar

   d. $\frac{3}{5}$ of a dollar

   e. $\frac{2}{3}$ of a dollar

   f. $3\frac{3}{4}$ dollars

   g. $2\frac{5}{6}$ dollars

   h. $1\frac{5}{8}$ dollars

   i. $9\frac{4}{5}$ dollars

   j. $12\frac{3}{20}$ dollars

4. Find a fraction that generates an equivalent calculator display for each of the following numbers. Then use your calculator to check your answers.

   a. 0.125

   b. 0.25

   c. 0.6

   d. 0.666666667

   e. 0.375

   f. 0.75

   g. 0.333333333

   h. 0.8

   i. 0.166666667

   j. 0.111111111

Name ______________________ Date ______________

Running for Class President (Page 1 of 2)

Four students are running for the office of class president at Cleveland Middle School. The elections will be held in two weeks. A survey of 600 students is conducted to determine which candidate is currently leading. The results of the survey are shown at the right.

Results of Class President Survey of 600 Students

Candidate	Number of Votes
Tom Cooper	99
Liza Varelli	204
José da Gamba	153
Lucia Candelo	73
Undecided	71

1. Explain whether or not each of the following statements about the survey results is accurate:

a. About $\frac{1}{6}$ of the students said that they will vote for Tom Cooper.

b. More than $\frac{1}{3}$ said that they will vote for Liza Varelli.

c. About $\frac{1}{5}$ said that they will vote for José da Gamba.

d. About $\frac{1}{8}$ said that they will vote for Lucia Candelo.

e. More than $\frac{1}{6}$ said that they do not know how they will vote.

f. About 33% of the students said that they will vote for Tom Cooper.

g. Fewer than 20% said that they will vote for Liza Varelli.

h. About 40% of the students said that they will vote for José da Gamba.

i. More than 10% said that they will vote for Lucia Candelo.

j. More than 20% said that they do not know how they will vote.

Name ____________________ Date ____________

Running for Class President (Page 2 of 2)

2. Use your calculator to determine the percent of students who will vote for each candidate and the percent of students who are still undecided.

3. Fill in the following pie chart to show the results of the survey. Choose a different color for each category and shade the legend and pie chart accordingly.

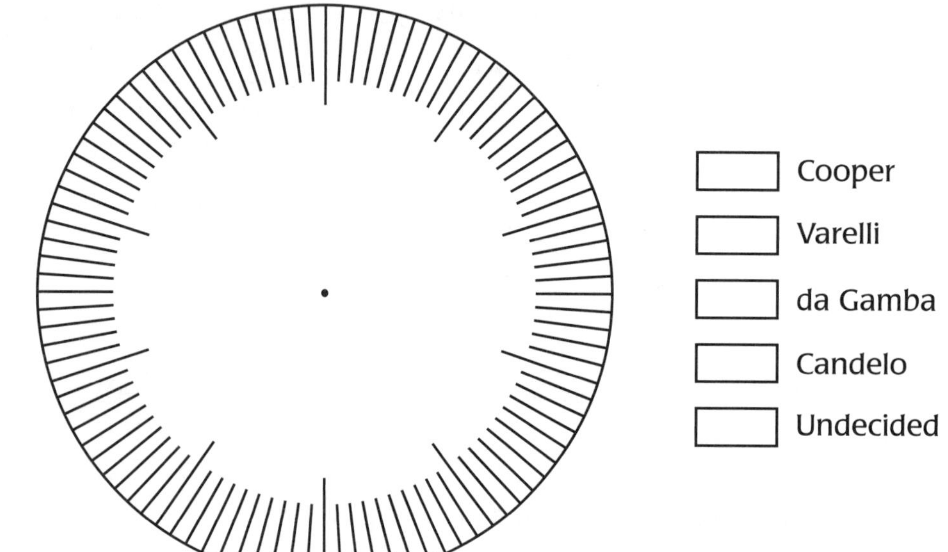

Name ______________________ Date ______________

Pie Chart Meter (Page 1 of 2)

The pie chart meter on the right is a circle whose outside ring is divided into 100 equal parts and whose inside ring is divided into 12 equal parts. You can use the pie chart meter to find equivalent expressions for fractions, decimals, or percents.

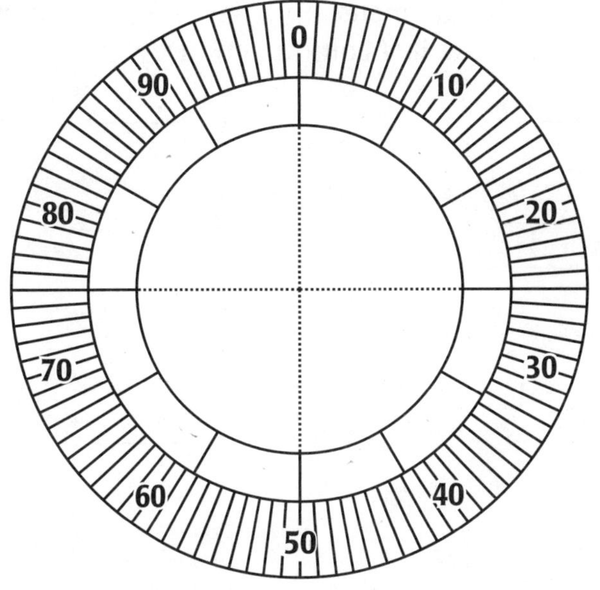

$\frac{1}{4}$ $\frac{2}{4}$ $\frac{5}{6}$ $\frac{4}{6}$ $\frac{4}{5}$ $\frac{75}{100}$ $\frac{1}{3}$ $\frac{2}{12}$ $\frac{6}{10}$ $\frac{5}{10}$ $\frac{10}{12}$ $\frac{8}{10}$ $\frac{3}{4}$ $\frac{40}{100}$ $\frac{3}{12}$ $\frac{4}{12}$ $\frac{2}{3}$ $\frac{3}{5}$ $\frac{1}{6}$ $\frac{2}{5}$

1. In the list above, each fraction has a corresponding equivalent fraction. Find the pairs of equivalent fractions. One pair is already identified for you below.

$\frac{1}{4} = \frac{3}{12}$ ______ = ______

______ = ______ ______ = ______

______ = ______ ______ = ______

______ = ______ ______ = ______

______ = ______ ______ = ______

Name ______________________ Date ______________

Pie Chart Meter (Page 2 of 2)

2. Use the pie chart meter on page 190 to convert the following fractions into equivalent percents. In the last column, choose your own fractions and convert them into percents.

$\frac{1}{2}$ = ______ %	$\frac{2}{3}$ = ______ %	______ = ______ %
$\frac{1}{3}$ = ______ %	$\frac{1}{12}$ = ______ %	______ = ______ %
$\frac{1}{6}$ = ______ %	$\frac{3}{4}$ = ______ %	______ = ______ %
$\frac{1}{4}$ = ______ %	$\frac{3}{8}$ = ______ %	______ = ______ %
$\frac{1}{8}$ = ______ %	$\frac{4}{5}$ = ______ %	______ = ______ %

3. Use the pie chart meter to determine if the following statements are true or false:

a. $\frac{2}{3}$ is a little less than 70%

b. $\frac{7}{12}$ is less than $\frac{3}{4}$

c. $\frac{1}{6}$ is half of $\frac{1}{3}$, so $\frac{1}{6}$ is more than 15%

d. $\frac{5}{6}$ is less than $\frac{60}{100}$

e. $\frac{7}{12}$ is less than 70%

Name ______________________ Date ______________

Fraction of a Fraction (Page 1 of 1)

1. One-half of the students in Ms. Abel's class are girls, and $\frac{1}{3}$ of the girls have brown hair. If there are four girls in the class who have brown hair, how many students are there in Ms. Abel's class?

2. One-third of the students in Mr. Tolme's class are boys, and $\frac{1}{2}$ of the boys play a musical instrument. How many boys in Mr. Tolme's class play a musical instrument? (*Note:* Many answers are possible.)

3. One-fourth of the students at Milton Middle School get a ride to school. One-half of these students ride with their parents. What fraction of the students ride with their parents?

4. Find the answer to each of the following:

   a. $\frac{1}{2}$ of $\frac{1}{4}$ **b.** $\frac{1}{3}$ of $\frac{1}{2}$ **c.** $\frac{1}{2}$ of $\frac{1}{3}$

   d. $\frac{1}{3}$ of $\frac{1}{3}$ **e.** $\frac{1}{2}$ of $\frac{1}{5}$ **f.** $\frac{1}{4}$ of $\frac{1}{3}$

   g. $\frac{1}{3}$ of $\frac{3}{4}$ **h.** $\frac{1}{2}$ of $\frac{2}{3}$ **i.** $\frac{3}{4}$ of $\frac{1}{2}$

   j. $\frac{2}{3}$ of $\frac{2}{3}$

Section K.
Fractions, Decimals and Percents

Name ______________________ Date ____________

Watermelons (Page 1 of 2)

Keith grows watermelons in his garden and then sells them for $1.80 per kilogram. Watermelons usually do not weigh exactly one or two kilograms, so Keith uses a double number line, as shown on the right, to find the price of each one.

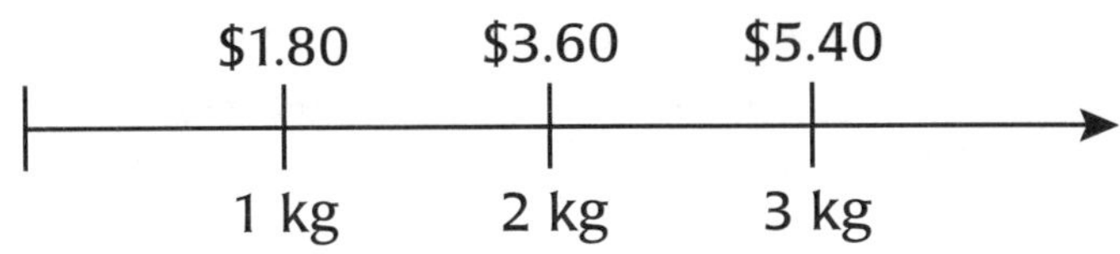

1. Explain how Keith finds the price of a watermelon weighing $1\frac{3}{4}$ kilograms by using the double number line.

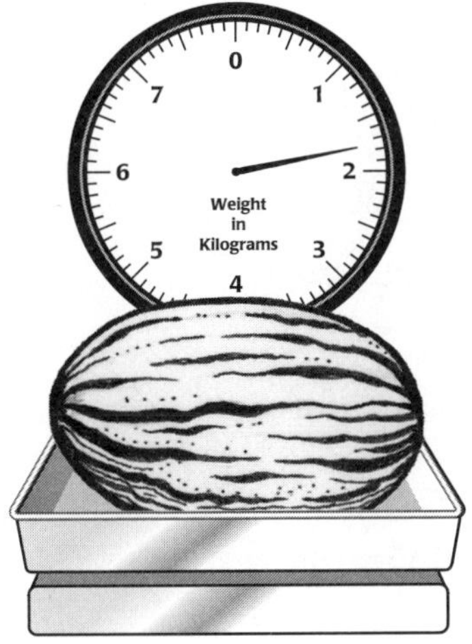

2. Use the double number line shown above to find the prices of watermelons that weigh the following amounts:

a. $\frac{3}{4}$ kilogram

b. $1\frac{1}{3}$ kilograms

c. $2\frac{2}{3}$ kilograms

d. $1\frac{1}{4}$ kilograms

e. $1\frac{1}{2}$ kilograms

Watermelons (Page 2 of 2)

3. Keith lowers his price for watermelons to \$1.20 per kilogram. What is the new price for watermelons that weigh the following amounts?

a. $\frac{3}{4}$ kilogram

b. $1\frac{1}{3}$ kilograms

c. $2\frac{2}{3}$ kilograms

d. $1\frac{1}{4}$ kilograms

e. $1\frac{1}{2}$ kilograms

4. Complete each of the following calculations:

a. $1\frac{3}{4} \times 80$

b. $2\frac{1}{4} \times 40$

c. $1\frac{1}{3} \times 180$

d. $2\frac{3}{4} \times 160$

e. $1\frac{2}{3} \times 360$

Name ______________________ Date ____________

At the Market (Page 1 of 3)

Jim runs a produce stand at the market. Since he is very good at math, Jim calculates the prices for his produce mentally, rather than using a calculator. To make calculations easier, he always rounds the amounts shown on the scale to the nearest benchmark fractions.

1. If one kilogram of grapes costs $1.60, find the prices that Jim will charge for the following amounts of grapes:

a.

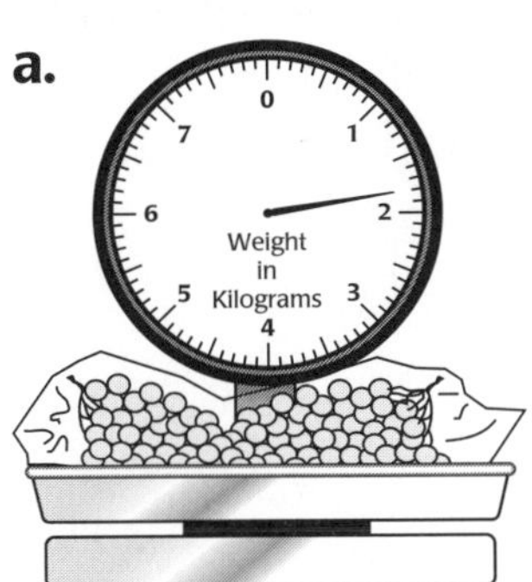

b.

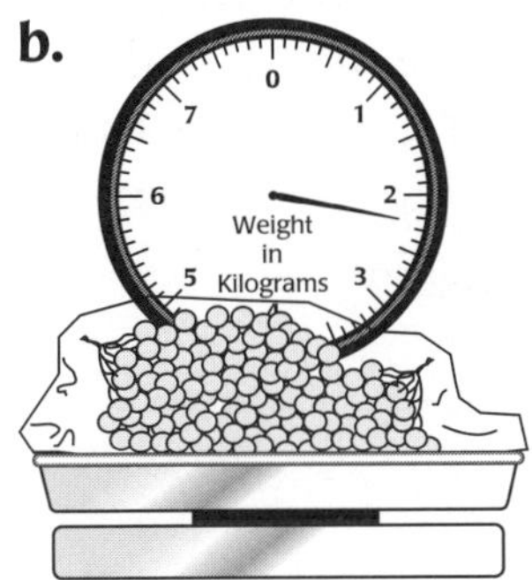

c.

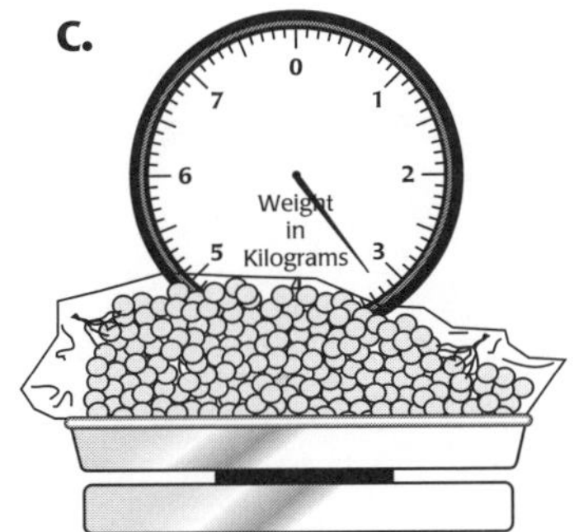

d.

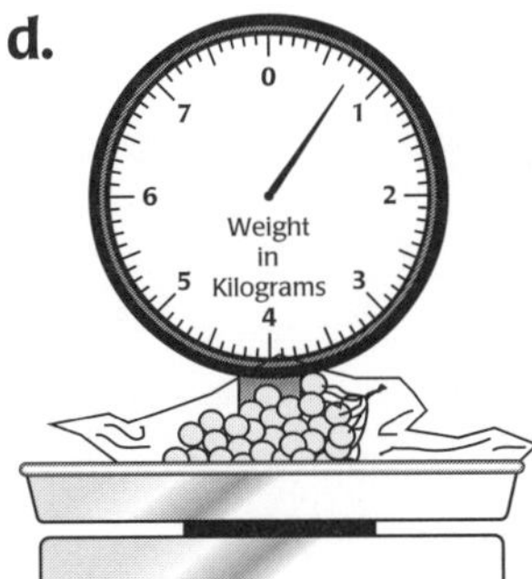

Name ______________________ Date ____________

At the Market (Page 2 of 3)

2. One day Jim's scale breaks down, so he borrows a digital scale that shows the weight of items in very precise decimal notation. For the first four items that Jim sells, the scale indicates the following weights: 2.212, 0.760, 1.461, and 2.110. Explain why the decimal scale might be more difficult for Jim to use.

3. Jim decides to convert the weights shown on the decimal scale to the nearest benchmark fraction. For example, if the scale shows 1.739, Jim converts this decimal to $1\frac{3}{4}$. Convert each of the following decimals to fractions that will be easy for Jim to work with mentally:

a. 2.772 kilograms

b. 3.236 kilograms

c. 1.401 kilograms

d. 2.352 kilograms

e. 2.534 kilograms

f. 1.690 kilograms

g. 2.110 kilograms

h. 3.728 kilograms

i. 0.317 kilogram

j. 2.289 kilograms

Name ______________________ Date ______________

At the Market (Page 3 of 3)

4. Find the price that Jim will charge for each of the following by first converting the decimal to a benchmark fraction and then using a mental calculation strategy:

   a. 1.327 kilograms of white grapes at $2.10 per kilogram

   b. 3.532 kilograms of apples at $1.10 per kilogram

   c. 0.728 kilogram of oranges at $4.00 per kilogram

   d. 0.229 kilogram of potatoes at $0.80 per kilogram

   e. 3.996 kilograms of watermelon at $1.80 per kilogram

Name ______________________ Date ______________

Calculator Calculations (Page 1 of 1)

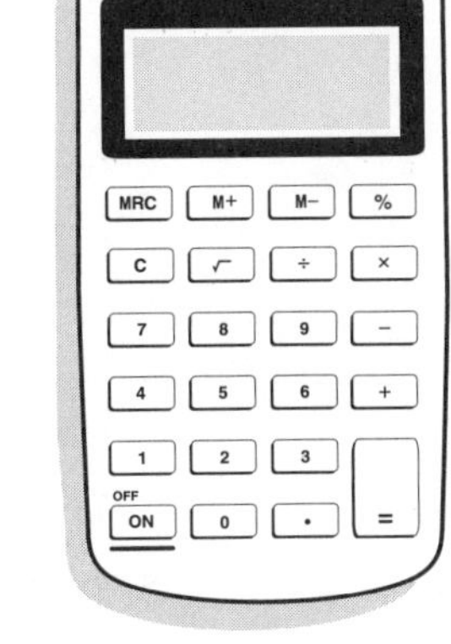

1. Use your calculator to find the cost of the following items. (*Note:* Stores round up all prices to the nearest cent.)

   a. 1.365 kilograms of pears at $3.10 per kilogram

   b. 0.723 kilogram of broccoli at $3.25 per kilogram

   c. 1.739 kilograms of collard greens at $1.79 per kilogram

   d. 1.396 kilograms of strawberries at $1.65 per kilogram

   e. 0.842 kilogram of oranges at $2.98 per kilogram

Michelle is supposed to use her calculator to do her homework, but the decimal point key is broken. To calculate the price of 1.293 kilograms of spinach that costs $2.98 per kilogram, Michelle enters 1293 × 298, and her calculator displays 385314. She can then figure out where to place the decimal point by estimating the correct answer.

2. Without using your calculator, find the answer to 1.293 × $2.98. Explain your strategy.

3. Without using your calculator, help Michelle find the correct price for each of the following items:

   a. 3.129 kilograms of grapes, selling for $3.10 per kilogram (Michelle's calculator displays 969990)

   b. 21.38 kilograms of apples, selling for $1.26 per kilogram (Michelle's calculator displays 269388)

   c. 0.729 kilogram of lemons, selling for $4.10 per kilogram (Michelle's calculator displays 298890)

   d. 3.28 kilograms of oranges, selling for $4.98 per kilogram (Michelle's calculator displays 163344)

   e. 0.083 kilogram of parsley, selling for $5.50 per kilogram (Michelle's calculator displays 45650)

Name ______________________ Date ______________

Sale (Page 1 of 1)

B & B Fashion is having a sale. Colored stickers indicate the percent discount: 10%, 15%, 20%, 25%, 33%, or 50%.

The price tag on a sweater is $48, and a sticker shows a discount of 25%. Angela and David each use a different way to calculate the resulting price.

Use either Angela's or David's method to calculate the sale price for each of the following items:

Item	Regular Price	Discount	Sale Price
Jacket	$96	25%	
Sweater	$48	$33\frac{1}{3}$%	
Jeans	$55	10%	
Shoes	$50	15%	
Winter Coat	$84	50%	
T-shirt	$8	20%	
Shorts	$32	25%	
Pants	$72	$33\frac{1}{3}$%	
Skirt	$82	10%	
Dress Shirt	$48	25%	

Name ______________________ Date ______________

Which Costs Less? (Page 1 of 1)

1. Soft Tunes and Audio Auction are both having sales. The stores feature the items listed below for the same regular price, but offer different discounts. Without using your calculator, determine which store has the lower sale price for each item.

Item	Regular Price	Soft Tunes Discount	Audio Auction Discount
CD Player	\$360	25% off	\$70 off
Portable Stereo/CD Player	\$270	$33\frac{1}{3}$% off	\$100 off
Speakers	\$548	20% off	\$100 off
Stereo Cabinet	\$598	15% off	\$100 off

2. Describe two ways to find 20% of \$450.

3. Calculate each of the following:

a. 20% of \$125

b. 25% of \$844

c. $33\frac{1}{3}$% of \$180

d. 10% of \$976

e. 15% of \$620

f. 25% of \$320

g. 10% of \$529

h. $66\frac{2}{3}$% of \$690

i. $33\frac{1}{3}$% of \$219

Section L. Ratios

Name ____________________ Date ____________________

How Fast? (Page 1 of 2)

1. Marge drove 65 miles from Springfield to Boville. She left Springfield at 2:00 P.M. and arrived in Boville at 3:15 P.M. She uses the ratio table shown on the right to find the average speed for her trip. Explain Marge's method.

Miles	65	260	52
Hours	$1\frac{1}{4}$	5	1

2. Instead of writing $1\frac{1}{4}$ hours for the travel time, you can use quarters of an hour or minutes. Show how to calculate the average speed for Marge's trip using the following ratio tables:

Miles	65			
Quarters of an Hour	5			

Miles	65			
Minutes	75			

Name ______________________ Date ______________

How Fast? (Page 2 of 2)

3. Use a ratio table to calculate the average speed for each of the following trips:

a. Departure Time: 8:00 A.M.; Arrival Time: 9:30 A.M.; Distance Traveled: 81 miles

Miles					

b. Departure Time: 2:00 P.M.; Arrival Time: 5:30 P.M.; Distance Traveled: 140 miles

Miles					

c. Departure Time: 8:15 A.M.; Arrival Time: 10:00 A.M.; Distance Traveled: 84 miles

Miles					

d. Departure Time: 9:05 A.M.; Arrival Time: 9:55 A.M.; Distance Traveled: 30 miles

Miles					

e. Departure Time: 7:30 A.M.; Arrival Time: 4:00 P.M.; Distance Traveled: 170 miles

Miles					

4. The average speed for the trip in part **e** above is very slow. Provide a possible explanation for the slow average speed.

Name ______________________ Date ____________

Gas Mileage (Page 1 of 2)

Mr. Van Dyke's science students are calculating the gas mileage of different cars. Gas mileage is the average number of miles that a car can travel on one gallon of gas.

1. Arnold is figuring out the gas mileage for a car that was driven 210 miles on 8.4 gallons of gasoline. He begins his calculations as shown below.

Miles	210	2,100			
Gallons of Gas	8.4	84			

a. Explain Arnold's first step.

b. Complete Arnold's calculations to find the car's gas mileage.

Name ______________________ Date ____________

Gas Mileage (Page 2 of 2)

2. Using ratio tables, calculate the gas mileage for each of the following:

a. A car travels 108 miles on 6 gallons of gasoline.

Miles					
Gallons of Gas					

b. A car travels 252 miles on 12 gallons of gasoline.

Miles					
Gallons of Gas					

c. A car travels 121 miles on 5.5 gallons of gasoline.

Miles					
Gallons of Gas					

d. A car travels 164 miles on 8 gallons of gasoline.

Miles					
Gallons of Gas					

e. A car travels 82.5 miles on 5.5 gallons of gasoline.

Miles					
Gallons of Gas					

3. If a car averages 20.5 miles per gallon of gas, how many gallons are needed to travel 492 miles? Use the following ratio table to calculate your answer.

Miles					
Gallons of Gas					

Name ______________________ Date ______________

Gas Mileage Again (Page 1 of 2)

Caroline's car averages 24 miles per gallon, which means that it needs 1 gallon of gasoline to travel 24 miles. Last year, Caroline drove her car 29,848 miles. She uses the following ratio table to calculate how many gallons of gas she used in her car during the year.

Gallons of Gas	1	1,000	200	40	1,240
Miles	24	24,000	4,800	960	29,760

Caroline concludes that she used a little more than 1,240 gallons of gasoline last year.

1. a. Explain Caroline's method.

   b. Show another way to use a ratio table to calculate the number of gallons of gasoline Caroline used in her car last year.

Name ______________________ Date ______________

Gas Mileage Again (Page 2 of 2)

2. For each of the following car trips, use a ratio table to compute the approximate number of gallons of gasoline used. Then check your work with a calculator.

a. Mr. Sommers drove 195 miles, and his car averages 24 miles per gallon.

Gallons of Gas						
Miles						

b. Ms. Donno drove 316 miles, and her car averages 18 miles per gallon.

Gallons of Gas						
Miles						

c. Ms. Bartok drove 428 miles, and her car averages 23 miles per gallon.

Gallons of Gas						
Miles						

d. Mr. Aspen drove 391 miles, and his car averages 22 miles per gallon.

Gallons of Gas						
Miles						

e. Mr. Yuanes drove his car a total of 19,362 miles last year, and his car averages 21 miles per gallon.

Gallons of Gas						
Miles						

Name ______________________ Date ______________

Horses (Page 1 of 2)

1. In Sun City, there is one horse per five inhabitants. How many horses are there per resident in Moon City?

2. In Dustown, there are 28 horses per 100 inhabitants.

a. Which town has relatively the smallest number of horses: Sun City, Moon City, or Dustown?

b. The population of Dustown is 1,368. How many horses are there in the town?

Name ______________________ Date ______________

Horses (Page 2 of 2)

3. Summarize the following information as a ratio (____ per ____).

 a. Switzerland has approximately 7,085,000 inhabitants and 2,362,000 television sets.

 b. Taiwan has approximately 21,501,000 inhabitants and 8,959,000 telephones.

 c. Greece has approximately 10,646,000 inhabitants and 4,259,000 radios.

 d. Japan has approximately 126 million inhabitants. Each day, approximately 72,668,000 newspapers are sold.

 e. Jamaica has approximately 2,574,000 inhabitants and 97,500 passenger cars.

Name ______________________ Date ______________

Cars (Page 1 of 2)

The table below shows the numbers of passenger cars and the population in several countries.

Country	Number of Passenger Cars	Population
Italy	28,200,000	58,262,000
Japan	39,000,000	125,506,000
Seychelles	4,700	73,000
United States	144,000,000	263,814,000
Zaire	105,000	44,061,000

1. a. Which country has the most passenger cars?

b. Which country has the most passenger cars relative to its population? Explain.

When comparing ratios, such as the ones above, it is helpful to organize your data in a ratio table. You can round numbers in ratio tables, as shown below.

Italy

Passenger Cars	28		
Population	58		

United States

Passenger Cars	144		
Population	264		

2. a. Explain what is shown in the two ratio tables above.

b. How can you determine which country has more cars relative to its population using the two ratio tables above?

Name ______________________ Date ______________

Cars (Page 2 of 2)

3. Order the five countries listed at the top of page 78 according to the number of cars relative to population.

4. You can express the relative number of cars in a country as the number of cars per 100 people. In Canada, for example, there are approximately 13.1 million passenger cars and 28.4 million people. This translates to 46 cars per 100 people.

a. Use a calculator to find the number of cars per 100 people for the five countries listed at the top of page 78.

b. Explain why it is easier to compare "cars per 100 people" than "cars per person" or "cars per 10,000 people" in part **a** above.

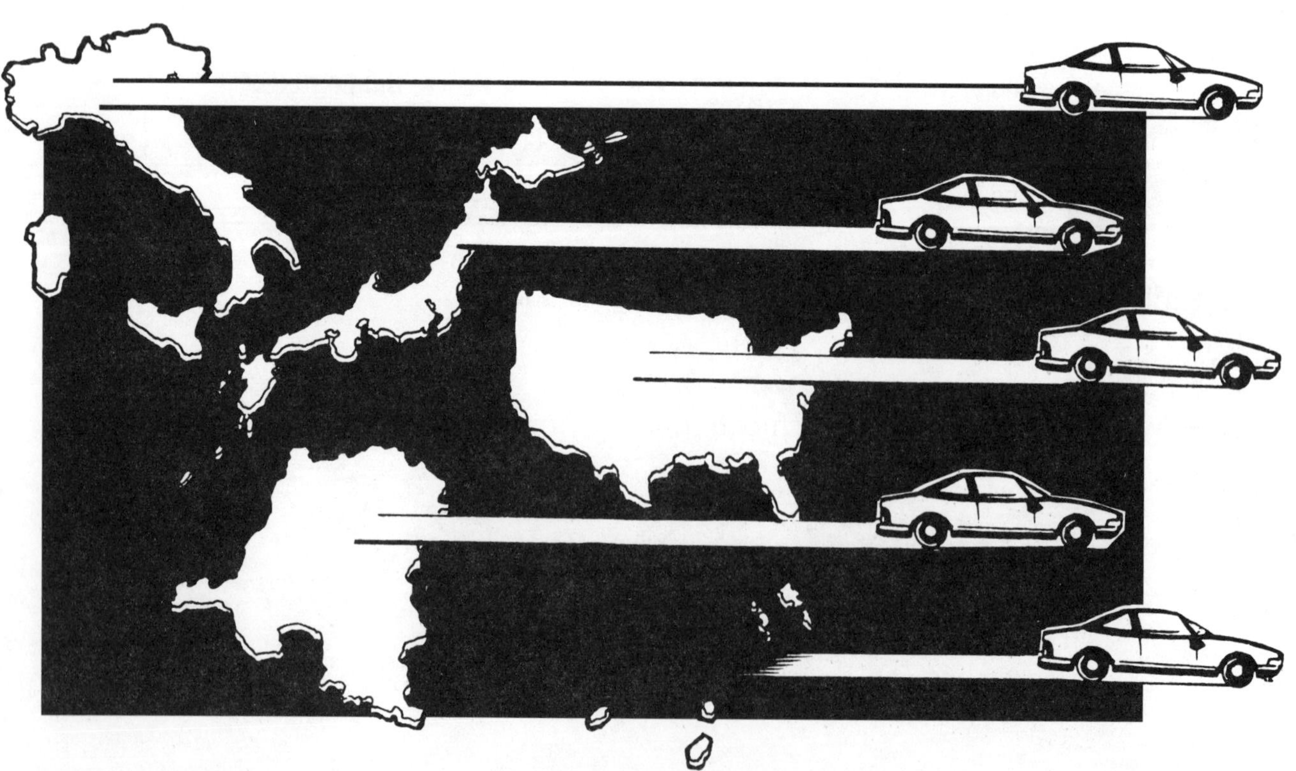

Name ______________________ Date ______________

From Ratios to Percents (Page 1 of 2)

The following methods can be used to convert a ratio into a percent:

- The Fraction Method: Rewrite the ratio as a fraction and then convert the fraction into a percent.
- The Ratio Method: Use a ratio table to calculate how many per 100.
- The Decimal Method: Use a calculator to express the ratio as a decimal by dividing and then convert the decimal into a percent.

1. Two out of five students have read *The Hobbit.* Use each of the methods described above to find a percent equivalent to two out of five.

2. Two out of three students have read *The Diary of Anne Frank.* Use each of the above three methods to determine a percent equivalent to two out of three.

Name ______________________ Date ______________

From Ratios to Percents (Page 2 of 2)

3. Find an equivalent percent for each of the following by using one of the methods described at the top of page 82:

a. Three out of 20 students participate in the drama club.

b. Seven out of 10 students have a bicycle.

c. Three out of eight students have read *The Outsiders.*

d. Three out of four students have read *A Wrinkle in Time.*

e. One out of three students has read *Roll of Thunder, Hear My Cry.*

f. Eight out of 12 students have seen the movie *Jurassic Park.*

g. Only one out of 12 students has read the book on which the movie *Jurassic Park* is based.

h. Five out of six students have been to the zoo.

i. The school has 250 students, but five are not in school today.

j. Mrs. Robinson's class has 14 girls and 11 boys.

Name ____________________ Date ____________

What Fraction? (Page 1 of 1)

1. The middle school in Woolton has 391 students. Fifty-one of the students are from Ridgeway, a small town nearby. The principal of the school claims, "One eighth of our students are from Ridgeway." Do you agree with the principal? Why or why not?

People sometimes round figures because precise amounts are not important in particular situations. Rounded figures are often better because they are easier to interpret and work with. Instead of saying "293 out of 391 students," for example, you can say "three-fourths of the students."

2. Use an easy-to-understand fraction to describe each of the following situations:

   a. Six-hundred fifteen readers have returned an opinion poll from *RadioWeek* magazine. Three-hundred ninety-seven say that WOLX is their favorite radio station.

   b. Of those 615 respondents, 125 like listening to country music.

   c. Fifty-nine of the 615 respondents regularly go to the movies.

   d. The population of Albertville is 54,273,372. Of all of the people in Albertville, 9,012,461 are children under the age of 10.

   e. In Dunhill, 9,321,989 people are under the age of 60. The population of Dunhill is 11,953,521.

   f. The school library has 937 books, of which 240 are fiction.

   g. There are 83 houses in Pilton, 63 of which are painted white.

   h. Yesterday Sondra counted her baseball cards and found that she had 78 of them. After giving some to a friend, she now has 61 cards.

   i. Juan had 52 baseball cards. After buying some more, he now has a total of 75 cards.